Design of an Integrated Push-Pull Tube Amplifier Made Easy

Giuseppe Amato

Copyright © 2018 Giuseppe Amato

All rights reserved.

ISBN-13: 9781719810500

Author can be contacted at:

giuseppe.amato@vtadiy.com

Book cover picture: courtesy of Sandro Amato

Rev. 1.5

Content:

CHAPTER 1: INTRODUCTION 1

1.1 What is in this book and what is not 2
1.2 How is this book organized? 3
1.2.1 Some difficult sections 4
1.3 Prerequisites 4
1.4 Acknowledgements 5
1.5 Apologies 5

CHAPTER 2: VACUUM TUBE BASICS 7

2.1 Hydraulic valves as a metaphor 7
2.1.1 Operating point 8
2.2 Electronic vacuum tubes 9
2.2.1 Diode 10
2.2.2 Triode 11
2.2.3 Tetrode 11
2.2.4 Tetrode in ultra-linear configuration 12
2.2.5 Pentode 13

CHAPTER 3: VACUUM TUBES AS AMPLIFIERS 15

3.1 Voltage and power amplifiers 15
3.2 The loadline 17
3.3 Operating point and bias 19
3.3.1 Linearity of the amplifier 21
3.4 Gain of the voltage amplifier 22
3.5 Amplifier classes 25
3.6 Biasing techniques 27
3.6.1 Fixed bias 27
3.6.2 Cathode bias or self-bias 29
3.6.3 Gain of the voltage amplifier with self-bias 31

CHAPTER 4: INTEGRATED PUSH-PULL VACUUM TUBE AMPLIFIER 37

4.1	OUTPUT STAGE OR POWER STAGE	**38**
4.1.1	SINGLE ENDED CONFIGURATION	38
4.1.2	IMPEDANCE OF AN OUTPUT TRANSFORMER	42
4.1.3	REACTIVE LOAD AND LOADLINE COMPUTATION	44
4.1.4	PUSH-PULL CONFIGURATION	46
4.1.5	PUSH-PULL ADVANTAGES	48
4.1.6	PUSH-PULL LOADLINE IN CLASS AB	51
4.2	PHASE SPLITTER STAGE	**53**
4.2.1	CONCERTINA PHASE SPLITTER	53
4.2.2	DC AND AC LOADLINE OF A CONCERTINA	56
4.2.3	GAIN OF THE CONCERTINA PHASE SPLITTER	57
4.3	INPUT STAGE	**58**
4.3.1	DIRECTLY COUPLED CONCERTINA	61
4.4	GLOBAL NEGATIVE FEEDBACK	**64**
4.4.1	GAIN WITH NEGATIVE FEEDBACK	67
4.4.2	BENEFITS OF NEGATIVE FEEDBACK	68
4.4.3	STABILITY OF NEGATIVE FEEDBACK	69

CHAPTER 5: POWER SUPPLY UNIT 73

5.1	POWER SUPPLY FOR THE AMPLIFIER STAGES	**73**
5.1.1	RECTIFIERS	74
5.1.2	RESERVOIR CAPACITOR	77
5.1.3	TRANSFORMER OUTPUT IMPEDANCE	79
5.1.4	LOAD ESTIMATION	80
5.1.5	ESTIMATION OF THE DC OUTPUT VOLTAGE	81
5.1.6	ESTIMATION OF THE RIPPLE VOLTAGE	82
5.1.7	ESTIMATION OF THE TRANSFORMER SECONDARY RMS CURRENT	83
5.1.8	SMOOTHING FILTERS	85
5.2	POWER SUPPLY FOR THE VACUUM TUBE FILAMENTS	**88**
5.2.1	IMPROVED CIRCUIT: COMMON AND ELEVATED VOLTAGE REFERENCE	90
5.2.2	ARTIFICIAL TRANSFORMER CENTRE TAP	92
5.2.3	REDUCING ELECTROMAGNETIC INTERFERENCES FROM OTHER TRANSFORMER WINDINGS 93	
5.3	POWER SUPPLY FOR THE FIXED GRID BIAS	**93**

5.3.1 FINE TUNING THE GRID BIAS 96
5.3.2 PROBING THE GRID BIAS CURRENT 98

CHAPTER 6: STEP BY STEP DESIGN OF A PUSH-PULL TUBE AMPLIFIER 101

6.1 DESIGN OF THE POWER STAGE 101
6.1.1 MAXIMUM OUTPUT POWER ESTIMATION AND VOLTAGE GAIN OF THE POWER STAGE 104
6.2 DESIGN OF THE INPUT AND PHASE SPLITTER STAGE 105
6.2.1 CONCERTINA PHASE SPLITTER 106
6.2.2 INPUT STAGE FOR DIRECTLY COUPLED CONCERTINA 108
6.3 GLOBAL NEGATIVE FEEDBACK 109
6.4 DESIGN OF THE AMPLIFIER POWER SUPPLY UNIT FOR THE AMPLIFIER STAGES ... 112
6.4.1 VOLTAGE DELIVERED BY THE RECTIFIER AND RIPPLE VOLTAGE 112
6.4.2 RATING OF THE POWER TRANSFORMER 114
6.4.3 RESISTORS OF THE SMOOTHING FILTERS 114
6.4.4 CAPACITORS OF THE SMOOTHING FILTERS 115
6.5 DESIGN OF THE VACUUM TUBE FILAMENT POWER SUPPLY 116
6.6 DESIGN OF THE FIXED GRID BIAS POWER SUPPLY 118

List of examples:

Example 1: Determining the loadline 18
Example 2: Gain of a voltage amplifier 25
Example 3: Determining the decoupling capacitor for fixed bias 28
Example 4: Determining the cathode resistor for self-bias 30
Example 5: Gain of a voltage amplifier with self-bias 35
Example 6: Inter-stage coupling capacitor of the power stage 42
Example 7: Impedance of an output transformer 43
Example 8: Determining the reactive loadline 45
Example 9: Biasing the Concertina phase splitter 55
Example 10: Inter-stage coupling capacitor in a concertina phase splitter 56
Example 11: DC and AC loadlines in a concertina phase splitter 57
Example 12: Loadline and bias of the input stage 59
Example 13: Coupling capacitor of the input stage 60
Example 14: Biasing for directly coupled Concertina 62
Example 15: Step network for negative feedback stability 70
Example 16: Power supply transformer output impedance 80
Example 17: Impedance offered by the amplifier to the power supply 81
Example 18: DC voltage output of a full wave rectifier 81
Example 19: Ripple voltage of a full wave rectifier 82
Example 20: RMS current in a transformer secondary of a full wave rectifier 84
Example 21: Determining resistor for a power supply RC Smoothing filter 86
Example 22: Determining voltage ripple of a power supply RC Smoothing filter 88
Example 23: Choosing heaters transformers 89
Example 24: Heater voltage elevation 91
Example 25: Power supply design for fixed grid bias 95
Example 26: Setting the grid bias voltage 97
Example 27: Balancing the grid bias voltage of two vacuum tubes in push-pull. 98
Example 28: Measuring the bias current 99

V

List of figures:

Figure 1: Vacuum tube as a hydraulic valve. 8
Figure 2: Operating point of a valve. 9
Figure 3: Diode and triode. 10
Figure 4: Tetrode. 12
Figure 5: Pentode. 13
Figure 6: Vacuum tube as a voltage amplifier. 16
Figure 7: Average anode characteristic graph and loadline. 19
Figure 8: Operating point of a vacuum tube voltage amplifier. 21
Figure 9: Equivalent circuit of the voltage amplifier. 24
Figure 10: Operating classes of an amplifier 26
Figure 11: Fixed bias schema. 29
Figure 12: Cathode or self-bias. 31
Figure 13: Equivalent circuit of the voltage amplifier with cathode resistor 33
Figure 14: Output gain attenuation with self-bias and bypass capacitor. 36
Figure 15: Integrated push-pull vacuum tube amplifier 38
Figure 16: Basic schema of a Single Ended power output stage. .. 39
Figure 17: Examples of various possible loadlines, for a power amplifier. 40
Figure 18: Impedance of real speakers 43
Figure 19: Reactive loadline. 45
Figure 20: Basic schema of a push-pull amplifier. 47
Figure 21: Dynamic behaviour of a push-pull amplifier. 48
Figure 22: Even order harmonic distortion in a push-pull amplifier. ... 50
Figure 23: Push-pull amplifier operating in class AB. 51
Figure 24: Loadline in a class AB push-pull amplifier. 53
Figure 25: Basic schema of a Concertina phase splitter. 54
Figure 26: DC and AC loadline of a concertina phase splitter 57
Figure 27: Basic schema of the input stage. 58
Figure 28: Loadline and operating point of the input stage. 61
Figure 29: Directly coupled Concertina. 62
Figure 30: Loadlines and operating points for directly coupled Concertina. 64
Figure 31: Basic schema for global negative feedback 66

Figure 32: Phase compensation for stable negative feedback............ 71
Figure 33: Phase compensation with a step network......................... 71
Figure 34: Basic components of a Power Supply Unit. 74
Figure 35: Half wave, full wave, and full wave bridge rectifiers. 76
Figure 36: The reservoir capacitor. 77
Figure 37: Ripple voltage reduction with reservoir capacitor............. 78
Figure 38: Transformer output resistance made explicit in an equivalent circuit. 80
Figure 39: Determining the output DC voltage of a full wave rectifier. 82
Figure 40: Determining the ripple voltage of a full wave rectifier. 83
Figure 41: Transformer secondary RMS current of a full wave rectifier. 84
Figure 42: Power supply smoothing filters. 87
Figure 43: Heater's power supply... 89
Figure 44: Heater's power supply elevated voltage reference 91
Figure 45: Power supply for grid bias 94
Figure 46: Improved schema for supplying the grid bias voltage 96
Figure 47: Probing the grid bias current 99
Figure 48: Schema of the amplifier output stage designed around EL84 tubes.. 102
Figure 49: Average anode characteristic graph and loadline of EL84 in ultra-linear configuration. 103
Figure 50: Schema and component values of the directly coupled input and concertina stages.. 106
Figure 51: Loadlines, operating points, and operating ranges for directly coupled concertina with 12AX7. 107
Figure 52: Schema and component values for global negative feedback 111
Figure 53: Power supply unit for the stages of the amplifier............. 112
Figure 54: Power supply for the filaments... 117
Figure 55: Power supply for the grid bias 120

Chapter 1: Introduction

Some time ago, a friend gave me his tube amplifier to test and compare against my solid-state amplifier. At the beginning, I was sceptical. However, I was immediately struck by the purity and clarity of the music it fed to my speakers. The more I used it, the more I realized that technology used fifty years ago was competing and even beating more recent technology.

I was even more surprised when my friend told me that his amplifier was a "Do It Yourself" project. I was impressed by the fact that it was possible to build at home an audio amplifier offering a comparable, or even better, audio quality than a branded, and supposedly good quality, amplifier.

Well, I took the challenge. For sure, on the internet I should have found plenty of good schemas and projects. I just needed to choose the best one to build my own first vacuum tube amplifier. I started searching and after a few clicks I collected a good number of options. It was very easy. At that point, I just needed to choose the best one. Here the problems started. Looking at the reviews of the various projects, both from supposed technical experts and audiophiles, I found conflicting opinions. The same project and solution were, at the same time, exceptional according to some and nonsense according to others.

In order to be able to discriminate I needed to learn about vacuum tube amplifier design and develop the needed skills to be able to judge. So, I started studying, simulating, prototyping, and measuring. Eventually, I was not only able to read and interpret projects. I was also able to design and build my first integrated push-pull vacuum tube amplifier.

Writing this book was motivated by the need to synthetize, in a simple and consistent form, all pieces of information that I collected. At the beginning, it was just for my own use. I wanted to avoid searching for information from scratch and draw conclusions I already figured out months before. I needed to preserve the information and knowledge I acquired for future use.

After a while I realized that this, with a small additional effort, could become a book for the general public, … well, … hopefully, … at least for the possibly interested public. This is how this book was born.

1.1 What is in this book and what is not

What I present in this book is not "the way" of designing vacuum tube amplifiers. It is just "my way" of designing an integrated push-pull vacuum tube amplifier. There are various other ways of designing and building vacuum tube amplifiers.

This book does not want to compete with several other books where an almost exhaustive discussion of the various ways of designing vacuum tube amplifiers is given. In this book, my objective is to propose a simple yet effective way of designing a vacuum tube amplifier. I start from the very basic concepts and follow a simple path that touches all needed aspects. Even if sometimes I need to go into formulas and details, I try to do it just when strictly needed.

For sure, there are some techniques that allow building better amplifiers, though with much higher complexity. However, it is also true that there are several other techniques, either simpler or more complex, resulting in much worse quality. It is easy to design an amplifier. What is not obvious is how to obtain an outstanding audio quality. The approach proposed in this book is simple and resulting quality is impressive as well. If you succeed in building your tube amplifiers according to my guidelines, I hope you can confirm this.

In this book, I do not address the issue related to actually building a vacuum tube amplifier. Therefore, I will not address issues related to soldering, wiring layout, placing components, grounding, assembling, etc.

In this book, I only address the issue of designing the schema of an integrated push-pull vacuum tube amplifier and its power supply unit. I discuss how to design the circuits, to choose the needed components, and to compute their values.

1.2 How is this book organized?

In this book, I made an extensive use of figures and proposed several practical examples, to design various parts of the circuits and to compute values of the needed components. I believe that this can significantly contribute to simplify some possible difficult concepts addressed.

The book first introduces some basic notions needed to design and understand vacuum tube amplifiers. Then, it considers all relevant aspects for designing an integrated vacuum tube amplifier, including its power supply unit. Finally, a real and concrete integrated push-pull vacuum tube amplifier is designed from scratch, using the various arguments discussed.

Specifically, Chapter 2: "Vacuum tube basics" introduces the needed concepts to understand vacuum tubes. This is a very basic introduction needed for the non-initiated. If you already know what vacuum tubes are and how they operate, you can skip this chapter.

Chapter 3: "Vacuum tubes as amplifiers" discusses how vacuum tubes can be used to obtain an amplifier. It introduces the notions of operating conditions, loadline, biasing techniques, and amplifier classes.

Chapter 4: "Integrated push-pull vacuum tube amplifier" is the core of this book. It goes into the details of designing an integrated push-pull vacuum tube amplifier. It discusses the single-ended and push-pull configurations, the various needed stages (power, phase-split, and input stages), and global negative feedback.

Chapter 5: "Power supply unit" discusses how to design the power supply unit for a vacuum tube amplifier. It introduces rectifier configurations and filters to reduce voltage ripple, to have a quiet amplifier. It explains how to estimate expected output DC voltage, ripple, and current delivered. It also discusses how to design power supply for the fixed bias circuit and for the filaments of the vacuum tubes.

Finally, Chapter 6: "Step by step design of a push-pull tube amplifier" uses all needed notions to design an entire integrated push-pull vacuum tube amplifier. It discusses the design of the power stage, the phase-split, and the input stage. It discusses the design of the global negative feedback loop. It also provides the design of the corresponding power supply unit.

1.2.1 Some difficult sections

I tried to use a simple language and to use equations and formulas just when needed. However, for completeness, I added some sections dealing with some advanced notions, which use non-trivial formulas and concepts. You can initially skip these parts, if you feel they are too complex. These difficult sections are:

- Section 3.4- "Gain of the voltage amplifier", where you learn how to estimate the gain of an amplifier.
- Section 3.6.3- "Gain of the voltage amplifier with self-bias" where you find a discussion on how to estimate gain of an amplifier under local negative feedback introduced by self-biasing.
- Section 4.2.3- "Gain of the concertina phase splitter" where gain of the concertina phase splitter is analysed.
- Sections 4.4.1-"Gain with negative feedback", 4.4.2-"Benefits of negative feedback", and 4.4.3-"Stability of negative feedback", where you find an analysis of the global feedback circuit, considering the resulting gain of the amplifier, the introduced benefits, and stability issues.

1.3 Prerequisites

There are very few prerequisites for an effective use of this book. You just need some basic notions of electronics. For instance, you need to know some basic equations like the Ohm's law, voltage divider, and low/high pass filters. You also need to know basic notions related to electric current, resistors, and capacitors.

1.4 Acknowledgements

A number of persons motivated, inspired, helped, and supported me during the preparation of this book. It is a pleasure and an honour to me to feel obliged to thank them.

I would like to start with Fabrizio, who first introduced me in the world of vacuum tube amplifiers. Sandro, whose meticulous attention to details prevented me from being too much inaccurate. When you write a book, you have an ideal reader in your mind. He is my ideal reader. Sandro also gave me the picture I used in the book cover.

I would like to especially thank my wife Giuseppina, for her patience, when I am completely captured by the enthusiasm of building vacuum tube amplifiers, and her inspirational support while writing this book. She is always positively unpredictable and capable of silently surprising me. I give a special acknowledge to my two sons Niccoló and Giacomo. They used to listen to music from their mobile phones and to make fun of me, when I was hypnotically listening to my music from my audio system. Eventually, they are able to distinguish between high- and low-quality audio systems and admit that listening to good music played by a good audio system is much better.

My mother Maria who, as a stereotypical Italian mum, is convinced and continuously tries to convince me as well, that I am the smartest, the cleverest, or simply said, the best. So, you would agree that this book is predestined to be the best book on vacuum tube amplifiers.

Last, I want to thank my dad Nicola. He, very recently, passed away. He taught me that there is not limit in what you can achieve when you are resolute. He introduced me to the philosophy of the "Do It Yourself", even if not applied to audio systems, and he silently transferred to me his enthusiasm and skills in finding solutions to any practical problems. Thanks dad!

1.5 Apologies

I am not a native English speaker. Probably, you might find some errors and some inaccurate sentences, in this book. I hope this does not significantly affect clarity, simplicity, and readability. I would appreciate if you can report me these errors so that I can correct them and improve the quality of the book.

In case, please send me an email at giuseppe.amato@vtadiy.com .

Chapter 2:
Vacuum tube basics

Readers, who already know what vacuum tubes are and how they operate, can skip this chapter and go directly to the next one. However, readers new to this field, before entering the details of designing a vacuum tube amplifier, need to learn some basics principles discussed in next sections.

2.1 Hydraulic valves as a metaphor

The operating modality of an electronic vacuum tube can be easily explained assimilating it to a *hydraulic valve*. Suppose a bucket, filled with water, is connected to a pipe. Suppose also that the pipe is connected to a hydraulic valve to control the flow of water. Refer to Figure 1, to follow the example. If the position of the bucket is higher than that of the other end of the pipe, the potential energy of the water, contained in the bucket, is higher than that of the end of the pipe. This implies that the water naturally flows from the bucket toward the bottom end of the pipe, producing a *current* of water.

However, if the valve is closed, as in Figure 1 a), there is no water flow. When the valve is fully open, as in Figure 1 b), the water flows freely. If the position of the valve is dynamically controlled with a signal, as sketched in Figure 1 c), the intensity of the water flow, through the pipe, follows the signal-controlled position of the valve.

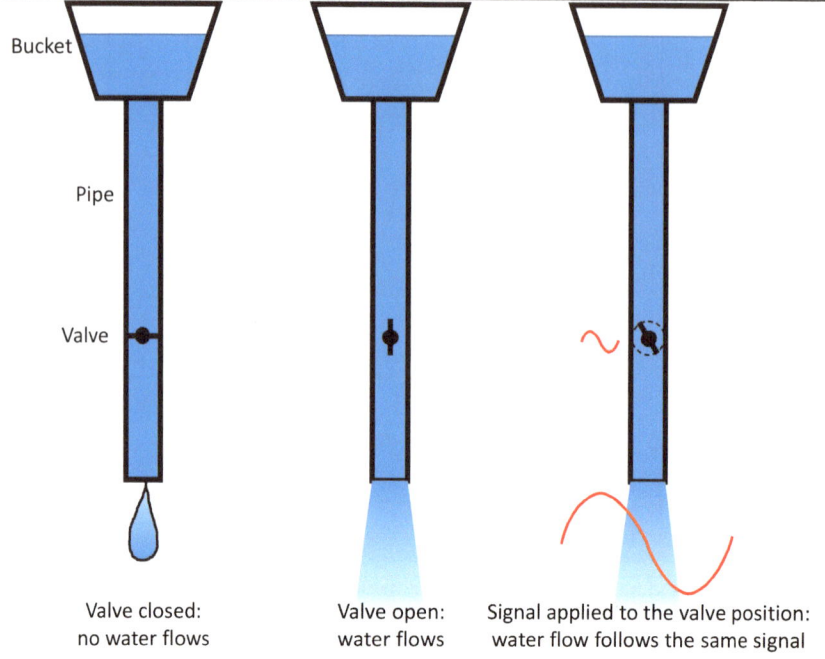

Figure 1: Vacuum tube as a hydraulic valve.
A vacuum tube can be assimilated to a hydraulic valve. When the valve is closed (a), no water flows through the pipe. When the valve is open (b) water flows. When position of the valve is dynamically controlled with a signal, the water flow follows the signal.

2.1.1 Operating point

Suppose an oscillating signal (the *input signal*) is applied to the valve position. The input signal has the effect of shifting the valve in a more open or less open position. The variation of the flow at the end of the pipe (*output or amplified signal*) follows the position of the valve.

In order to correctly amplify the input signal, the position of the valve, in correspondence of the no-signal state, needs to be carefully chosen. This is the quiescent position of the valve. Suppose that the quiescent position of the valve is set so that, when there is no signal the valve is fully closed. In this case, the flow variation cannot follow the negative part of the input signal. In fact, when the input signal goes below the quiescent state, since the valve is already fully closed, no further reduction of the flow is possible. Similarly, if the valve is fully open at the quiescent state, it is not possible to follow the positive part of the signal. When the input signal is above the quiescent state, no further increase of the flow is possible, since the valve is already fully open.

The quiescent position of the valve is referred as the *operating point*. In a few words, it is the position of the valve when no signal is applied to it. Alternatively, we can consider the operating point as the amount of water that flows when no signal is applied to the valve.

In general, when the operating point is too high, as shown in Figure 2 a) the highest parts of the signal is lost. When the operating point is too low, as shown in Figure 2 b) the lowest parts of the signal is lost. In Figure 2 c) the entire dynamic range of the signal is correctly amplified.

We will see that choosing the appropriate operating point, of an electronic vacuum tube, is a very relevant aspect

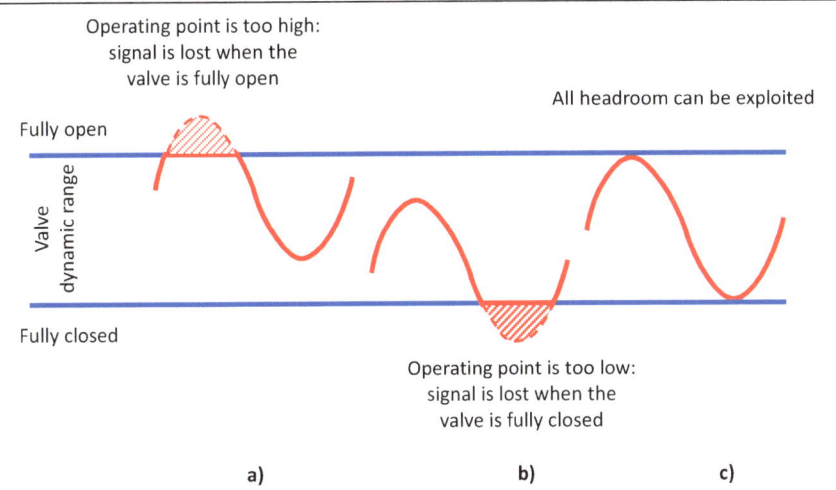

Figure 2: Operating point of a valve.
If the operating point is too high as in a), signals that need to turn the valve in a position higher than the fully open positions, are lost. If the operating point is too low as in b), signals that need to turn the valve in a position that is lower than the fully close position, are lost. If the operating point is correctly set, the full dynamic range of the signal can be correctly amplified.

2.2 Electronic vacuum tubes

The operating principle of an *electronic vacuum tube* is similar to that of a hydraulic valve. Electrons travel along electric wires or vacuum space, in place of water flowing in pipes. The electric potential (or voltage)

forces electrons to traverse electric wires and vacuum space, in place of the potential energy due to the effect of the gravity pushing water to flow through pipes. Finally, an electronic valve can control the flow of electrons, analogously to what described above for hydraulic valves.

In the following sections, we will introduce the basic types of electronic vacuum tubes and their operating principles.

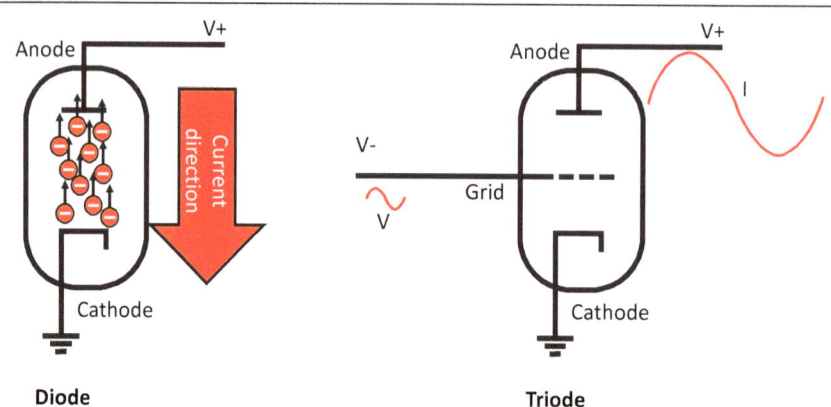

Figure 3: Diode and triode.
Diode, on the left. Triode on the right. When the cathode of a vacuum tube is heated, electrons are emitted. If the anode has an electric potential higher than the cathode, the emitted electrons are attracted by it and an electric current start to flow (by convention the current goes from the positive to the negative end, even if the electrons, which have a negative charge, move from the negative to the positive end). In a triode, the negative electric potential of the grid, with respect to the cathode, can be used to control the flow of electrons from the cathode to the anode.

2.2.1 Diode

The simplest type of electronic vacuum tube is the *diode*, depicted on the left of Figure 3. A diode has two terminals: a *cathode* and an *anode*. When the cathode is sufficiently heated, electrons start to leave its surface, due to the *thermionic effect*. When an electric potential V+, higher than that of the cathode, is applied to the anode, the electrons emitted by the cathode, which have a negative charge, are attracted and move toward the anode itself.

By convention, the direction of the electric current is that of the "positive" charges. Therefore, even if the electrons move from the cathode to the anode, we say that the electric current goes from the anode to the cathode. In fact, a current of negative charges in one direction is equivalent to a current of positive charges in the opposite direction.

2.2.2 Triode

What we have just described above is very similar to the example we made in Section 2.1, where the water flows from the bucket through the pipe under the effect of the gravity. Here the electric current flows from the high potential V+, applied to the anode, toward the ground potential of the cathode. However, a valve that allows controlling this flow is still needed.

In a *triode*, as depicted on the right of Figure 3, a third terminal, called *grid*, is placed between the anode and the cathode. When the electric potential of the grid is lower than that of the cathode, the electrons emitted by the cathode are repelled and find difficulties to reach the anode. If the grid is negative enough, current is blocked, obtaining the same effect of closing the valve in the hydraulic circuit. If the grid potential is the same than that of the cathode, the current flows freely from the anode to the cathode, obtaining the same effect of fully opening the valve in the hydraulic circuit. Intermediate negative grid voltages modulate the current flowing from the anode to the cathode[1]. When a voltage signal is applied to the grid, the electric current, from the anode to the cathode, follows the signal applied to the grid.

It is important to note that no current traverses the grid in normal operations. Being the grid negative, with respect to the cathode, it repels electrons and there is no electric current.

2.2.3 Tetrode

The basic idea of the triode was refined with the introduction of the *tetrode*. A tetrode, as shown in Figure 4 on the left, has a fourth electrode called *screen*, between the anode and the grid. The screen has the

---

[1] Note that the grid potential can also be positive with respect to the cathode. In this case, also the grid starts attracting the electrons emitted by the cathode, and current flows from the grid to the cathode as well. This situation is very dangerous, since the vacuum tube can be damaged. However, this effect is sometime exploited in guitar amplifiers in order to explicitly produce distorted sound. In this book, we will not take into consideration this possibility, since we are discussing high fidelity audio amplifiers, where distortion must be avoided.

purpose of reducing the capacitance produced by the grid and the anode. In fact, in a triode, the grid and the anode are very close one to the other and jointly behave as a small capacitor, which might cause instability and oscillations. If the screen has a voltage higher than that of the cathode and the grid, but lower than the anode, it acts as an electrostatic screen between the grid and the anode, thus reducing their inherent capacitance.

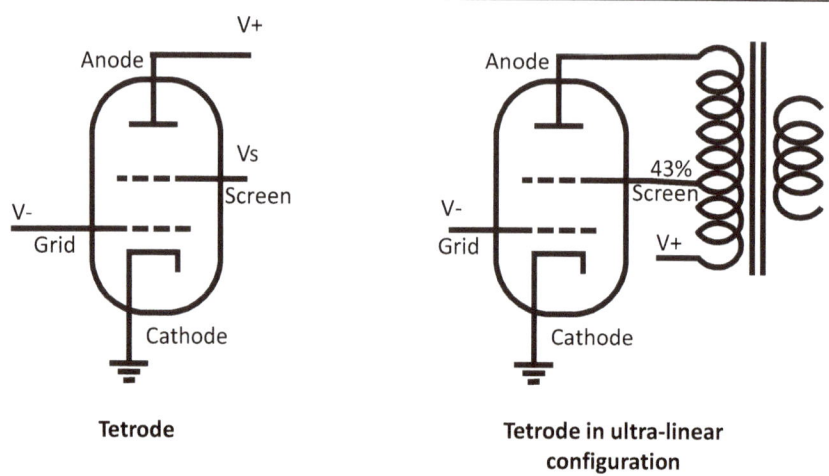

Figure 4: Tetrode.
In a tetrode, on the left, a fourth electrode is used as a screen to limit the inherent capacitance between grid and anode. The screen can also be set to operate in ultra-linear configuration, on the right, by providing it with a percentage of the anode output signal.

2.2.4 Tetrode in ultra-linear configuration

In a tetrode, the screen is positive with respect to the cathode. Thus, it attracts a certain amount of the electrons emitted from the cathode itself, which would have gone to the anode. The result is a small current flowing through the screen. This effect is exploited to configure a vacuum tube to work in *distributed load*, or *ultra-linear* modality. This configuration is obtained by feeding back a percentage of the anode output signal to the screen, rather than applying a fixed voltage to it. The wanted percentage of anode output signal is typically provided to the screen by connecting it to a tap, coming out from the output transformer, as shown on the right side of Figure 4. The current through the screen, produces a sort of negative feedback and, with the appropriate percentage of anode signal, distortion falls at very small values, just slightly reducing power efficiency. The optimal percentage to be applied to the screen depends on the specific electronic vacuum tubes used. In many power amplifier designs, this percentage is generally set around 43%.

2.2.5 Pentode

Tetrodes were further refined with the introduction of the *pentode*. When electrons emitted by the cathode reach the anode, they might have enough energy to stimulate secondary electron emission from the anode itself. Secondary emission electrons might reach the screen grid, causing instability and oscillations. In order to prevent secondary emission electrons from reaching the screen grid, *pentodes* use a fifth electrode, called *suppressor*, as shown in Figure 5. The suppressor is generally directly connected to the cathode, either with an internal connection in the vacuum tube, or by an explicit connection between the corresponding pins, as depicted by the dashed line in Figure 5.

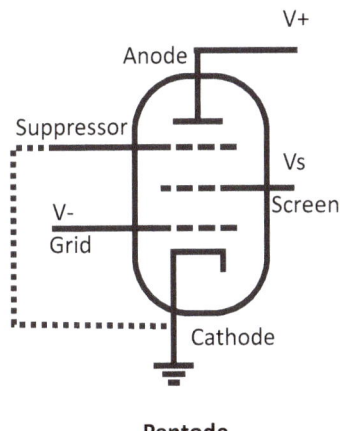

Pentode

Figure 5: Pentode.
In a pentode, a fifth electrode, called suppressor is added to attract secondary emission electrons, emitted by the anode when hit by electrons arriving from the cathode. This prevents secondary emission electrons from reaching the screen grid, causing instability and oscillations.

Chapter 3:
Vacuum tubes as amplifiers

The variation of the voltage, applied to the grid of an electronic vacuum tube, produces a variation of the current from the anode to the cathode. In vacuum tubes used in input stages (for instance 12AX7 vacuum tubes), current is generally limited in a range from 0 up to a few milliamperes. In vacuum tubes used in power stages (for instance EL34 vacuum tubes), current ranges from 0 up to a few hundreds of milliamperes.

With an appropriate configuration, this current variation can be exploited to obtain voltage and/or power amplifiers, where the voltage signal applied to the grid is conveniently amplified.

3.1 Voltage and power amplifiers

Figure 6 shows a very common configuration where a resistor (*load*) R_L is placed between the power supply V+ and the anode. The load and the resistance presented by the vacuum tube itself form a voltage divider.

This configuration composes a basic *voltage amplifier*, where the output signal is taken between the anode and the load. In fact, when the voltage grid varies, the current traversing the vacuum tube varies as well, because of the variation of the vacuum tube resistance. Consequently, the voltage measured between the anode and the cathode, called the *anode voltage* or the *plate voltage*, changes according to the voltage divider equation:

$$V_a = V_+ \frac{R_V}{R_V + R_L},$$

where V_a is the anode voltage, R_v, is the resistance offered by the vacuum tube, regulated by the grid voltage, and R_L is the load resistance. In an appropriately designed voltage amplifier, small grid voltage variations ΔV_g produce much larger anode voltage variations ΔV_a.

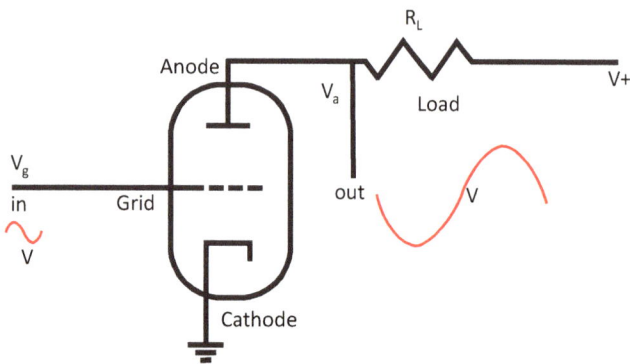

Triode as a voltage amplifier

Figure 6: Vacuum tube as a voltage amplifier.
A vacuum tube can be used as a voltage amplifier by placing a load between the anode power supply and the anode itself. The load forms a voltage divider with the resistance offered by the vacuum tube, which can be controlled by varying the grid voltage. The voltage signal applied to the grid is amplified by the voltage signal seen at the anode.

Note that *the amplified voltage signal is inverted* with respect to the signal voltage applied to the grid. In fact, when the signal voltage increases, the current from anode to cathode also increases, that is the resistance offered by the vacuum tube decreases, and correspondingly the voltage seen at the anode decreases as well.

Amplifiers with a large voltage gain, and limited current variation, are *voltage amplifiers*. They are generally used to increase the input signal before passing it to a power amplifier.

Amplifiers offering also a significant current variation are *power amplifiers*. They are generally used where high power is required across the load. In this case, the load is not just used to form a voltage divider with the vacuum tube. The load is the component that actually makes use of the amplified power. In power amplifiers built using vacuum tubes, as discussed later, generally the load consists of an output transformer, which adapts the impedance of the speaker to the load required by the vacuum tube.

In case of vacuum tubes used in input stages, as for instance 12AX7 vacuum tubes, given that the current variation from the anode to the cathode is very small, only voltage amplification is generally exploited.

In case of an output/power vacuum tube, as for instance EL34 tubes, given that significant current variation occurs from the anode to the cathode, we have a power amplifier.

3.2 The loadline

A very useful information, generally provided in the datasheet of an electronic vacuum tube, is the *average anode characteristics graph*. Figure 7 shows this graph for a 12AX7 vacuum tube. In the graph, each black plot is obtained by maintaining fixed the grid voltage V_g, and varying the anode voltage V_a. Each black plot is respectively tagged with the corresponding grid voltage V_g used. The graph shows, for a fixed grid voltage V_g, the relationships between the anode voltage V_a and the anode current I_a. When V_a increases, I_a increases as well along the plot corresponding to the chosen V_g.

In a voltage amplifier, as the one in Figure 6, once the power supply $V+$ is fixed, when the grid voltage V_g varies, the vacuum tube resistance varies and V_a and I_a vary as well, as discussed before. The anode voltage V_a and the current I_a vary together along a straight line, as for instance the green line in Figure 7. This line is called the *loadline* and is very useful for setting up the vacuum tube operating point, choosing the power supply voltage, and the load of the amplifier. The slope of the loadline depends on the load resistance R_L. Its intersection, with the anode voltage axis, depends on the power supply $V+$.

The loadline, in case of a resistive load, can be drawn by connecting together the point corresponding to maximum anode voltage and minimum conduction, with the point corresponding to minimum anode voltage and maximum conduction, as discussed in the following example.

> **Example 1: Determining the loadline**
>
> Let us suppose the power supply is $V+=300V$. As an example, let us first consider the (non realistic) case where no load is applied to the anode, that is when $R_L=0$. In this case, the anode voltage is always equal to the power supply voltage $V_a=V+$. When $V_g=-4.0$, the corresponding plot in the graph indicates that at $V_a=300V$ there is no conduction, since $I_a=0$. If we vary the grid voltage from -4.0V to -1.5V, given that the anode voltage is always $V_a=300V$, the current varies from 0 up to 3.15 mA, following the red vertical line, drawn in the figure. In fact, the plot corresponding to $V_g=-1.5V$ indicates that when $V_a=300V$, we have $I_a=3.15$ mA.
>
> Suppose now that the load applied between the anode and power supply is $R_L=150K$ Ohms. Let us first determine the point of minimum conduction. When the vacuum tube does not conduce, the tube internal resistance is infinite $R_v=\infty$. Therefore, the voltage seen at the anode is $V_a=300V$ and current is zero. Let us now suppose that the vacuum tube does not offer any resistance. In this case, the anode voltage V_a is zero[2] and the voltage drops from 300V to 0 through the load. The anode current is obtained using the Ohm's law as
>
> $$I = \frac{V_+}{R_L} = \frac{300V}{150kOhm} = 2.0 mA .$$
>
> Accordingly, the loadline intersects the horizontal axis at 300V and the vertical axis at 2.0 mA, obtaining the green line in Figure 7.
>
> Grid voltage variations between 0V and -4V shift anode voltage and current along this line.

A load with a small impedance produces a more vertical loadline. Large load impedance makes a more horizontal loadline. Small impedance

---

[2] Note that this condition cannot be reached in practice. In fact, a vacuum tube always offers an internal resistance which prevents the anode voltage from going to 0.

produces more current and less voltage swing, while large impedance mainly produces voltage variations.

Resuming, the loadline indicates what happens at the anode of a vacuum tube used as a voltage amplifier, when a negative voltage is applied to its grid. A signal applied to the grid is seen as a movement along the loadline.

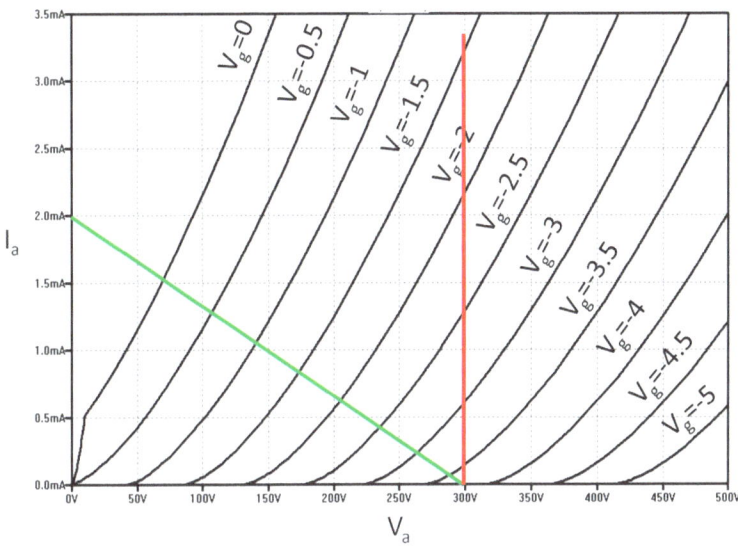

Figure 7: Average anode characteristic graph and loadline.
Every black plot corresponds to the relationship between anode voltage (V_a) and current from anode to cathode (I_a), at a fixed grid voltage. If the anode voltage is maintained fixed and the grid voltage varies, the current varies as well following a vertical line (for example along the red line). If a load is applied between the anode and the power supply, this forms a voltage divider with the vacuum tube and both voltage and current vary when the grid voltage varies (for example along the green line). The line along which the vacuum tube operates depends on the impedance of the load and is called the loadline.

3.3 Operating point and bias

The *operating point* of a vacuum tube is the position on the loadline, corresponding to the voltage and current measured at the anode in the quiescent status, that is when no signal is applied to the grid. The operating point is set by providing the grid with an appropriate fixed negative voltage V_g, with respect to the cathode. This fixed voltage is commonly referred as the *grid bias* voltage, or simply as the *bias*, of the vacuum tube.

Using the average anode characteristic graph and the loadline, the operating point is identified by the intersection of the loadline with the plot corresponding to the chosen grid bias voltage V_g. Let us use Figure 8 and the green loadline to exemplify this. Suppose that the grid bias voltage V_g is set to -1,5V. In this case, the operating point is the red spot, at the intersection between the green loadline and the plot, corresponding to a grid voltage V_g of -1.5V. This corresponds to a quiescent anode voltage V_q of 175V and a quiescent anode current of 0.85 mA.

Suppose now that an input signal V_{in} ranging between +1 and -1V is applied to the grid. The voltage seen at the grid is V_g+V_{in} and ranges between -0.5 and -2.5V. This produces a corresponding oscillation along the loadline, around the operating point, as depicted by the red range in the figure. The voltage measured at the anode is V_q+V_{out} and oscillates, roughly, in the range between 110V and 240V, that is V_{out} ranges from -65 to +65V.

If the operating point is set too high, that is if the grid bias voltage is too high, as exemplified by the blue spot and range, then *saturation* would occur when the signal oscillation produces a grid voltage above 0V. In fact, when grid voltage is zero, the vacuum tube has the maximum conduction, and no additional current is possible. On the other hand, if the bias is set too low, as depicted by the violet spot and range, when the voltage seen at the grid is too low, the vacuum tube would quit conducting.

Note that, as better discussed in Section 3.5, there are cases where it is not needed to amplify the entire input signal and the operating point is set on purpose so that just a portion of the signal is correctly amplified.

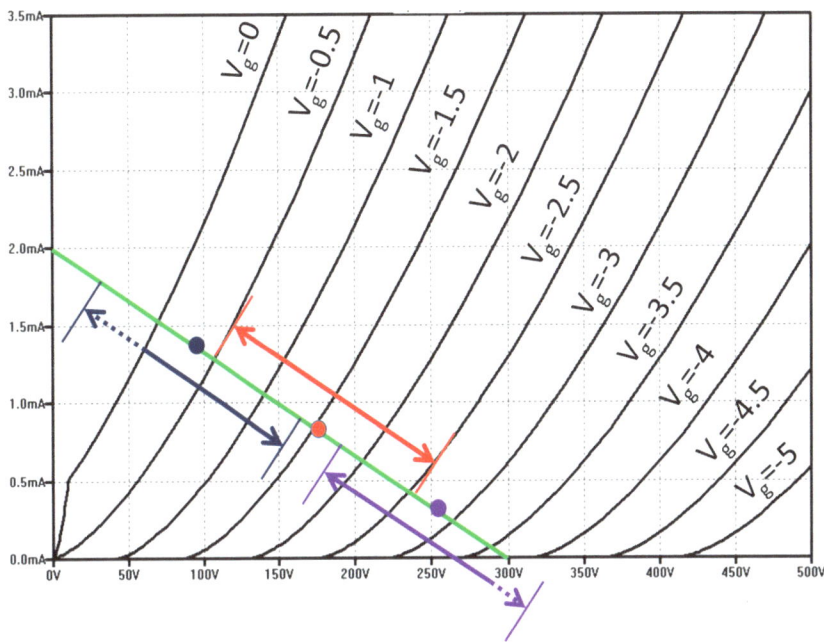

Figure 8: Operating point of a vacuum tube voltage amplifier.
If the operating point is too high (blue spot and blue range), large signals applied to the grid produce saturation, since when the grid voltage is 0 we have the maximum conduction between cathode and anode. On the other hand, when the operating point is too low (violet spot and violet range), too small signals bring to zero conduction between cathode and anode. The operating point also determines the linearity response of the vacuum tube. If the operating point is correctly chosen (for instance, red spot and red range), the vacuum tube responds linearly to the entire voltage range applied to the grid.

3.3.1 Linearity of the amplifier

The optimal choice of the loadline, and the quiescent operating point, should identify a range where the vacuum tube operates linearly. Linear operation means that variations of the voltage applied to the grid should be linearly reflected in variations along the loadline. Graphically, this means that the black plots, corresponding to different grid voltages, should be uniformly distributed along the portion of the loadline used. For instance, in Figure 8, the intersections between the black plots and the green loadline are denser in correspondence of value of the grid voltage around -3.0V, rather than voltages around -1.0V. This means that, when the grid voltage varies around -3.0V, it produces variation of the anode voltage smaller than when the grid voltage varies around -1.0V. If the grid is provided with a signal that spans from -1.0V to -3.0V, this is not linearly amplified and harmonic distortions are generated.

3.4 Gain of the voltage amplifier

The *gain A* of a voltage amplifier is the ratio between the output signal and the input signal. More formally, we have:

$$A = \frac{V_{out}}{V_{in}}$$

Sometimes it is convenient to express the gain in dB as follows:

$$A_{db} = 20 \cdot \log(A).$$

The gain of a voltage amplifier depends on the specific characteristics of the vacuum tube used, and on the load resistance. Datasheets of electronic vacuum tubes generally specify some useful parameters to determine how the signal applied to the grid is amplified. Two very useful parameters are the *amplification factor* μ and the *anode resistance* (or *plate resistance*) r_a.

The *amplification factor* is defined as the ratio between the anode voltage variation and the grid voltage variation, when the anode current is maintained constant:

$$\mu = -\frac{\Delta V_a}{\Delta V_g}, \text{ with } I_a \text{ constant.}$$

The amplification factor represents the maximum possible voltage amplification for a vacuum tube, which can be ideally (but not practically) obtained using a load with infinite resistance. This parameter is very stable on various operating conditions of the vacuum tube. The minus sign indicates that the anode voltage phase is inverted with respect to the grid voltage phase.

The other useful parameter is the *anode resistance*. It is defined as ratio between the anode voltage variation and the anode current variation, when the grid voltage is maintained constant:

$$\boxed{r_a = \frac{\Delta V_a}{\Delta I_a}}, \text{ with } V_g \text{ constant.}$$

The anode resistance is the resistance offered by the vacuum tube when the grid voltage is maintained constant and the anode voltage varies. It can be represented as a series resistance between the anode of the vacuum tube and the anode pin. Its value is not stable and depends on the specific operating conditions. Vacuum tubes datasheets, generally, report various values of the anode resistance, in correspondence of various anode voltages. If the datasheet does not report the anode resistance for the chosen configuration, it can be obtained from the anode characteristic graph, using previous equation.

When a load is connected to the anode, the real gain of the amplifier is smaller than the amplification factor μ. In fact, as long as the loadline becomes more vertical, the anode voltage variation, in correspondence of the same grid voltage variation, becomes smaller. For instance, previously we discussed that when the load resistance is zero the loadline is vertical, so no voltage amplification occurs. The real voltage gain A of the voltage amplifier, which is the ratio between the anode voltage variation and the grid voltage variation, in presence of a load, can be computed using the amplification factor μ, the anode resistance r_a, and the value of the load resistance R_L.

Let V_g be the grid bias voltage and V_q the corresponding quiescent anode voltage. When the AC input signal V_{in} is added to the grid bias voltage V_g, the anode voltage becomes $V_q + V_{out}$. The value of V_{out} can be determined using the equivalent circuit in Figure 9, where the vacuum tube is replaced by an AC power supply. In the equivalent circuit, the grid bias, the quiescent anode voltage, and the power supply V+ are not considered, given that they are constant. The voltage of the AC power supply (which replaces the vacuum tube) is equal to the amplification factor times the input voltage $-\mu V_{in}$. The minus sign here indicates that the phase of the AC power supply is reversed with respect to that of V_{in}. The output amplified signal V_{out} is taken from the voltage divider formed by the anode resistance r_a and the load resistance R_L. Its value is obtained using the voltage divider equation as

$$V_{out} = -\mu V_{in} \frac{R_L}{R_L + r_a}.$$

Finally, using the equation for the voltage gain we have:

$$A = \frac{\Delta V_{out}}{\Delta V_{in}} = -\mu \frac{R_L}{R_L + r_a}.$$

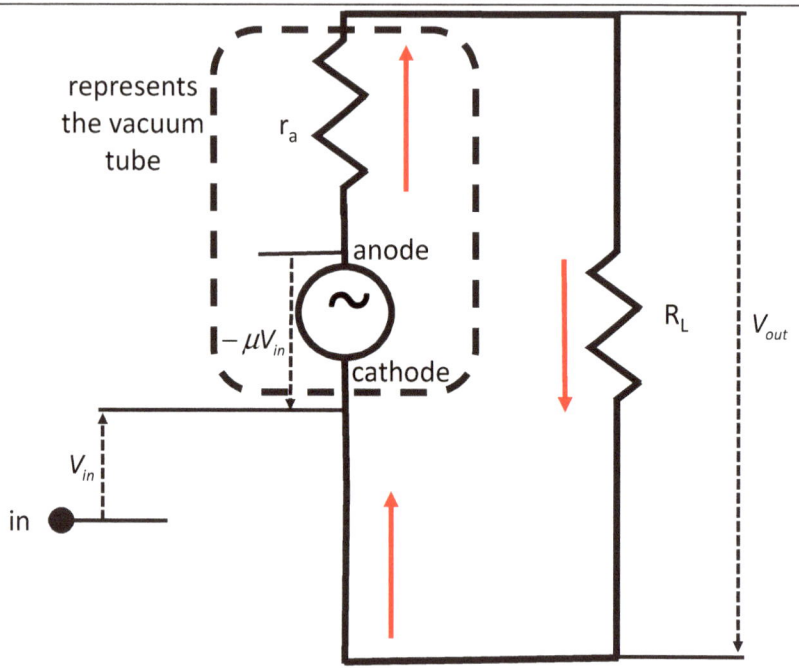

Figure 9: Equivalent circuit of the voltage amplifier.
The vacuum tube is seen as an AC voltage source equal to the amplification factor μ times the input signal V_{in}, with opposite phase. The anode resistance and the load resistance are in series to the voltage source to form a voltage divider. The output amplified signal V_{out} is taken at the voltage divider.

Generally, the above equation is reported without the minus sign, given that we are mainly interested in the amplitude of the signal, rather than its phase. So, we have the following equation for the voltage gain of an amplifier:

$$\boxed{A = \mu \frac{R_L}{R_L + r_a}}$$

Example 2: Gain of a voltage amplifier

Consider for instance a 12AX7 vacuum tube and suppose it is configured according to the green loadline in Figure 8, corresponding to a load of 150K Ohms. Suppose the operating point is set with a grid bias V_g of -1.5V, corresponding to an anode voltage V_a of 175V.

The 12AX7 datasheet specifies an amplification factor $\mu=100$. The anode resistance at the anode voltage of 175V, in correspondence of a grid bias of -1.5V, can be computed as follows.

We take, from the anode characteristic graph, the currents in correspondence of voltages slightly lower and higher than 175V, along the plot corresponding to -1.5V grid voltage. Then, we use anode resistance equation to determine it. The current in correspondence of an anode voltage of 190V is 1.0 mA. The current in correspondence of an anode voltage of 160V is 0.6 mA. Using the anode resistance equation, we have:

$$r_a = \frac{\Delta V_a}{\Delta I_a} = \frac{190V - 160V}{1.0mA - 0.6mA} = 75kOhm.$$

We now have all we need to estimate the gain of this voltage amplifier:

$$A = \mu \frac{R_L}{R_L + r_a} = 100 \frac{150k}{150k + 75k} = 66.6.$$

3.5 Amplifier classes

The operating point also determines the operating class of an amplifier.

In a *Class A* amplifier, the grid bias is set so that the vacuum tube conducts and amplifies the entire input signal. For instance, if the operating point is set in correspondence of the red spot in Figure 8, the obtained amplifier operates in class A, since it can amplify the entire input signal. The plot at the top of Figure 10 shows a signal amplified by a Class A amplifier. In a Class A amplifier, significant current traverses the vacuum tube also in the quiescent state, that is when no input signal is provided to the grid.

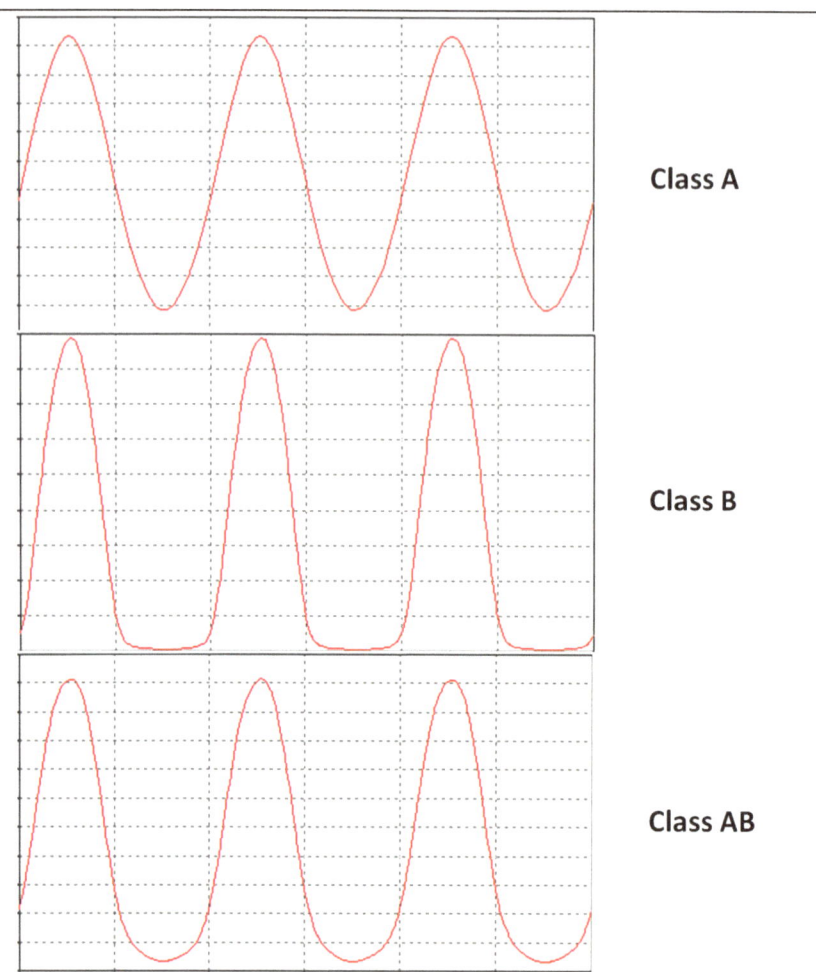

Figure 10: Operating classes of an amplifier.
Top plot depicts Class A operation, where the entire signal is amplified. Centre plot shows Class B operation. In this case, just half signal is amplified. Class AB amplifiers, in the lower plot, amplify most of the signal, and cut-off when the input signal is below a certain threshold.

A *Class B* amplifier amplifies just half input signal. Suppose the bias is set so that it is very close to the intersection of the green loadline and the horizontal axis in Figure 8. In this case, just the upper part of the signal is amplified, and the vacuum tube quits conducting in correspondence of the remaining part of the signal. The plot in the middle of Figure 10, shows a signal amplified by a Class B amplifier. Class B amplifiers are more efficient, since there is no current traversing the vacuum tube when there is no input signal.

Class AB amplifiers are a compromise between class A and Class B. The bias is set so that most of, but not all, the signal is amplified. Suppose the operating point is set in correspondence of the violet spot in Figure 8. In this case, the upper part of the signal is entirely amplified. The lower part of the signal is amplified until the intersection of the green loadline and the horizontal axis is reached. At this point the vacuum tube quits conducting, so part of the output signal is missing. Lower plot in Figure 10, depicts a signal amplified by a Class AB amplifier. Class AB amplifiers are more efficient than Class A amplifiers and less than Class B amplifiers, since just a small current traverses the tube when no signal is amplified.

3.6 Biasing techniques

We discussed that a grid voltage, negative with respect to the cathode voltage, controls the current that traverses a vacuum tube and that the grid bias voltage determines the quiescent vacuum tube operating point. The input voltage signal is added to the grid bias voltage and is amplified.

There are two largely used techniques to provide a vacuum tube with a grid bias voltage, negative with respect to the cathode. The *fixed bias* technique requires a separate power supply that provides the wanted negative voltage. The *cathode bias* or *self-bias* technique connects the grid to ground and elevate the cathode voltage above ground. In this way the grid voltage is negative with respect to the cathode voltage. Let us discuss these two biasing techniques more in details.

3.6.1 Fixed bias

The *fixed bias* schema is given in Figure 11. Two different power supplies are used. PS_1 gives the high-tension *V+* to the anode of the vacuum tube, through the load. The negative of PS_1 is connected to ground. PS_2 produces the needed grid bias voltage V_g. The negative grid bias $-V_g$ is obtained by connecting the positive of PS_2 to ground. In this way, the positive of PS_2 is at ground level and the negative is at $-V_g$ with respect to ground. The grid receives the $-V_g$ bias voltage using a resistor R_l. Given that no current goes through the grid, in normal operations, the resistor R_l does not affect the voltage seen by the grid. The cathode is also

27

connected to ground so that the grid is at $-V_g$ with respect to the cathode, as needed.

The usage of resistor R_l will be better discussed in Section 4.1.1. For the moment, we just mention that one of the purposes of R_l (also called the *grid leak*) is to provide the input signal, received from previous stage, with a high impedance path to ground.

The capacitor C_d, from the grid leak resistor to ground, decouples the residual input signal, that traverses R_l, from the bias voltage supply. Consider that, generally the bias power supply provides bias voltage to several vacuum tubes in the amplifier. For instance, in a stereo amplifier, both left and right channels are sometimes biased by the same power supply. The residual input signal, which traverses R_l, is added to the bias voltage and goes also to the other channels, where it is amplified by the other vacuum tubes, creating problems of cross-talk. In order to avoid that, the capacitor C_d forms, with the resistor R_l, a low-pass filter that shorts to ground the residual input signal. The value of this capacitor should be large, so that even very low frequencies are shorted to ground and do not go to the grid of the other vacuum tubes.

Example 3: Determining the decoupling capacitor for fixed bias

Suppose, for simplicity, that R_l is the only resistance seen by the capacitor C_d, and suppose its value is 200K Ohms. In order to have a low-pass filter with a very low cut-off frequency, for instance 1 Hz, we use the low-pass filter equation and we obtain the following capacitance:

$$C_d = \frac{1}{2\pi \cdot R_l \cdot f} = \frac{1}{2\pi \cdot 200KOhm \cdot 1Hz} = 0.8 \mu F.$$

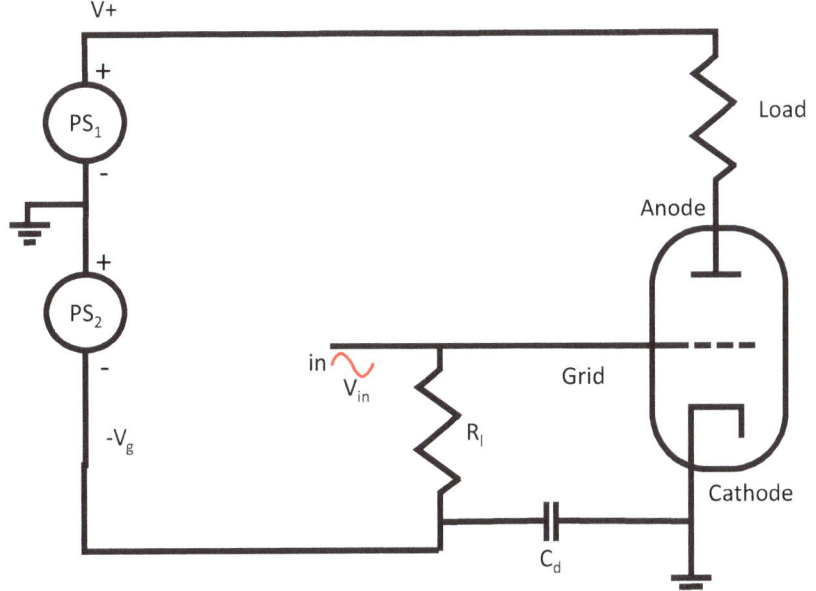

Figure 11: Fixed bias schema.
Fixed bias can be obtained by using two separate power supplies. PS_1 produce the high-tension V+ to be given to the vacuum tube anode. PS_2 produces the voltage V_g for the grid. The negative voltage $-V_g$ is obtained by connecting the positive of PS_2 to ground. Also the negative of PS_1 is connected to ground. In this way, the negative of PS_1 and the positive of PS_2 are at ground voltage, the negative of PS_2 is at $-V_g$ and is connected to the grid.

3.6.2 Cathode bias or self-bias

Negative voltage between grid and cathode can also be obtained by connecting the grid to ground voltage and by elevating the cathode voltage. This technique is generally referred as *cathode bias* or *self-bias*. The cathode voltage is elevated by connecting it to ground through the resistor R_k, generally called the *cathode resistor*, as shown in Figure 12. Given that, generally, there is an anode current also at the quiescent state, the resistor R_k produces a voltage drop from the cathode to ground so that the cathode voltage is above ground. The grid, being at ground voltage, is negative with respect to the cathode.

Note that, also in this case, the grid is not directly connected to ground. Rather, a grid leak resistor R_l is used to provide the input signal with a high impedance path to ground, as we already discussed for the fixed bias. Since there is no current flowing through the grid, it is at ground voltage.

The value of R_k can be computed using the Ohm's law by knowing the *bias current*, that is the cathode current at the operating point (quiescent state), as shown in next example.

Example 4: Determining the cathode resistor for self-bias

Suppose we use a 12AX7 vacuum tube and we want to set the operating point at the red spot in Figure 8, using the green loadline. This corresponds to a bias current of 0.75 mA. Using the average anode characteristics graphs, we see that a grid voltage of -1.5V, with respect to the cathode, is needed to obtain a current of 0,75 mA. Using the Ohm's law, we find that the resistance needed to elevate the cathode at 1.5V, when the current is 0.75 mA, is:

$$R_k = \frac{V}{I} = \frac{1.5V}{0.75mA} = 2kohm.$$

It is important to mention that the cathode resistor introduces a form of local negative feedback. In fact, when the current increases, the cathode voltage increases as well. In this case, the grid becomes more negative, with respect to the cathode, and tends to reduce the vacuum tube conduction. When current decrease, we have the opposite effect and the grid becomes less negative, increasing vacuum tube conduction. In other words, the cathode resistor tends to oppose the amplification of the signal and reduces the gain of the vacuum tube. In order to mitigate and almost eliminate this effect, a *bypass or decoupling capacitor C_k*, is generally introduced in the circuit, as shown in Figure 12. The bypass capacitor compensates the cathode voltage variation trying to maintain it as stable as possible, when amplifying a signal. In this way, local negative feedback is significantly reduced and gain significantly increased, as discussed in next Section.

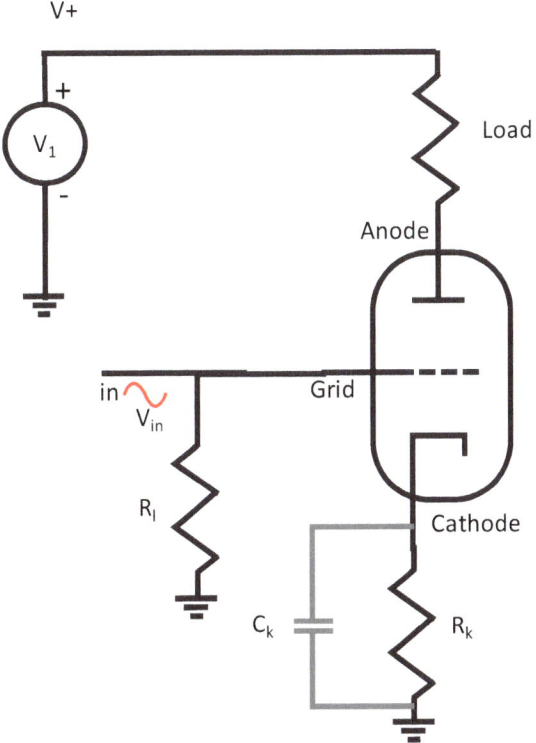

Figure 12: Cathode or self-bias.
The negative voltage of the grid, with respect to the cathode, can be obtained by connecting the grid to ground voltage and by elevating the cathode voltage by means of a resistor called the cathode resistor. By computing the resistor value, according to the wanted bias current, the cathode voltage is elevated so that the grid has the correct negative bias voltage. The cathode resistor can also be bypassed by a bypass capacitor to reduce the negative feedback, introduced by the cathode resistor, and increasing gain.

3.6.3 Gain of the voltage amplifier with self-bias

The cathode resistor, used for self-bias, introduces a form of local negative feedback: when the current increases, the cathode voltage increases as well reducing the grid to cathode voltage, and vice versa. The result is that the cathode resistor reduces the gain of the vacuum tube. This local negative feedback can be significantly neutralized by using a bypass capacitor connected in parallel to the cathode resistor.

The gain of the voltage amplifier with self-bias can be determined using the equivalent circuit shown in Figure 13. Let us first suppose that no bypass capacitor C_k is used. Similarly, to what we discussed in section 3.4, when we estimated the gain of the voltage amplifier, the vacuum tube is represented by an AC power supply, in series with the anode resistance r_a. Let V_{in} be the voltage of the input signal measured from the grid to the ground, that is from the grid to the terminal of the cathode resistor R_k opposite to the cathode. The voltage V_{fb} measured between the two terminals of R_k, corresponding to the voltage drop introduced by R_k itself, is the feedback voltage. The grid to cathode voltage V_{in}^{fb}, resulting from the combined action of the input and feedback voltages, is:

$$V_{in}^{fb} = V_{in} + V_{fb}.$$

The above equation can be better understood using the equivalent circuit in Figure 13, as we already did in Section 3.4. In the circuit, power is supplied by the AC power supply, replacing the vacuum tube. Voltage drops, through the resistors in the circuit, proceeding clockwise. Suppose highest voltage is at the top of the AC power supply (the anode of the vacuum tube), and lowest voltage is at the bottom of the AC power supply (the cathode of the vacuum tube). V_{in} is the voltage difference between the input *in* and the cathode resistor end, opposite to the voltage source (opposite to the cathode). The voltage difference between the input *in* and the other cathode resistor end (between the grid and the cathode) is higher than V_{in}, because of the voltage drop introduced by R_k. Given that the voltage drop is V_{fb}, V_{in}^{fb} will be equal to V_{in} plus V_{fb}.

The AC power supply produces a voltage equal to $-\mu \cdot V_{in}^{fb}$. As we said in Section 3.4, the minus sign here indicates that the phase of the AC power supply is reversed with respect to that of V_{in}^{fb}. The output voltage V_{out}^{fb} is taken between the two ends of the load R_L and can be computed using the voltage divider equation as

$$V_{out}^{fb} = -\mu \cdot V_{in}^{fb} \frac{R_L}{r_a + R_L + R_k}.$$

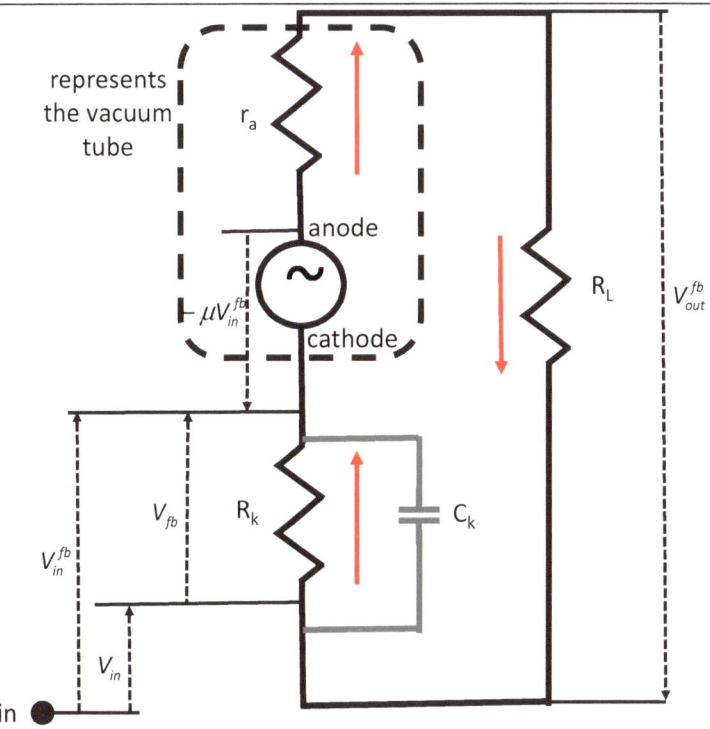

Figure 13: Equivalent circuit of the voltage amplifier with cathode resistor
The AC power supply produce a voltage equal to the amplification factor μ times the voltage between the input (grid of the vacuum tube) and the lower extreme of the voltage source itself. The minus sign indicates that the phase is reversed, with respect to the input signal. Each resistor introduces a voltage drop along the circuit, in the direction of the red arrow. The output signal is taken at the extremes of the load resistor R_L. To neutralize the local negative feedback, a capacitor C_k can be connected in parallel to R_k.

Using the voltage divider equation again, the feedback voltage V_{fb} is

$$V_{fb} = -\mu \cdot V_{in}^{fb} \frac{R_k}{r_a + R_L + R_k}.$$

Replacing V_{fb} into the equation for V_{in}^{fb} and simplifying we have:

$$V_{in}^{fb} = V_{in} + V_{fb} = V_{in} \frac{r_a + R_L + R_k}{r_a + R_L + R_k(1+\mu)}.$$

Now we can replace this into the equation for V_{out}^{fb} and we obtain:

$$V_{out}^{fb} = -\mu \cdot V_{in} \frac{R_L}{r_a + R_L + R_k(1+\mu)}.$$

Finally, we can express the gain of the voltage amplifier, with local feedback introduced by a non-bypassed cathode resistor, as:

$$\boxed{A^{fb} = \frac{V_{out}^{fb}}{V_{in}} = \mu \cdot \frac{R_L}{r_a + R_L + R_k(1+\mu)}.}$$

As we said in Section 3.4, we do not consider the phase so we omit the minus sign.

Let us now consider the case where a capacitor C_k is used to bypass the cathode resistor R_k, as shown in Figure 13. The impedance of the capacitor depends on the signal frequency f and is

$$Z_c = \frac{1}{2 \cdot \pi \cdot f \cdot C_k}.$$

The impedance of the cathode resistor in parallel with the bypass capacitor is

$$Z_k = \frac{R_k \cdot Z_c}{\sqrt{R_k^2 + Z_c^2}}.$$

The gain of the voltage amplifier, with local feedback introduced by a cathode resistor and bypassed by a capacitor, is obtained by replacing R_k with Z_k in the equation for A^{fb}. We obtain:

$$\boxed{A_{bypsd}^{fb} = \mu \cdot \frac{R_L}{r_a + R_L + Z_k(1+\mu)}.}$$

The value of the gain A_{bypsd}^{fb} depends on the frequency and on the capacitor. It ranges between these two extremes:

$$\mu \cdot \frac{R_L}{r_a + R_L + R_k(1+\mu)} \leq A_{bypsd}^{fb} \leq \mu \cdot \frac{R_L}{r_a + R_L}.$$

Minimum gain, equal to the non-bypassed cathode resistor, occurs when the capacitor impedance is maximum (at very low frequencies and/or very low capacitances). Maximum gain, similar to the circuit without cathode resistor, occurs when the capacitor impedance is minimum (at high frequencies and/or large capacitance).

The value of the capacitor should be chosen so that gain is maximum even at very low audible frequencies, as discussed in next example.

Example 5: Gain of a voltage amplifier with self-bias

Let us consider again the configuration of a voltage amplifier with the green loadline in Figure 8. In Example 1, we set the load R_L to 150K Ohms. In Example 2 we determined that the anode resistance r_a is 75K Ohms, and the gain of the amplifier at the operating point identified by the red spot, with no cathode resistor, is 66.6. In Example 4, we determined that the cathode resistor needed to set the operating point at the red spot, using self-bias, is 2K Ohms.

Remember that he amplification factor of a 12AX7 vacuum tube is μ=100. Therefore, the gain of the voltage amplifier, using self-bias with this cathode resistor is

$$A^{fb} = 100 \cdot \frac{150 KOhm}{75 KOhm + 150 KOhm + 2 KOhm(1+100)} = 35.12.$$

Considering that the full gain with no cathode resistor is 66.6, the new gain corresponds to 20·log(35.12/66.6) = -5.5dB with respect to full gain.

Suppose now we use a bypass capacitor of 150µF. At a frequency of f=1 Hz the capacitor impedance is

$$Z_c = \frac{1}{2 \cdot \pi \cdot 1 \cdot 150 \mu F} = 1.061 KOhm.$$

The impedance of the cathode resistor in parallel with the bypass capacitor is

$$Z_k = \frac{2 KOhm \cdot 1.061 KOhm}{\sqrt{2^2 KOhm + 1.061^2 KOhm}} = 937 Ohm.$$

The gain of the voltage amplifier at 1 Hz, with the bypass capacitor, is:

$$A_{bypsd}^{fb} = 100 \cdot \frac{150KOhm}{75KOhm + 150KOhm + 937Ohm(1+100)} = 47.$$

This corresponds to 20·log(47/66.6) = -3dB with respect to full gain.

Repeating the same process for f=10Hz, we obtain a gain of 63.6, corresponding to 20·log(63.6/66.6) = -0.4dB. This is fairly acceptable, since no attenuation is realistically perceived at all audible frequencies.

Figure 14 reports the output gain attenuation, with the above configuration, with frequency ranging up to 100 Hz. It can be seen that the attenuation is practically eliminated from 40Hz on.

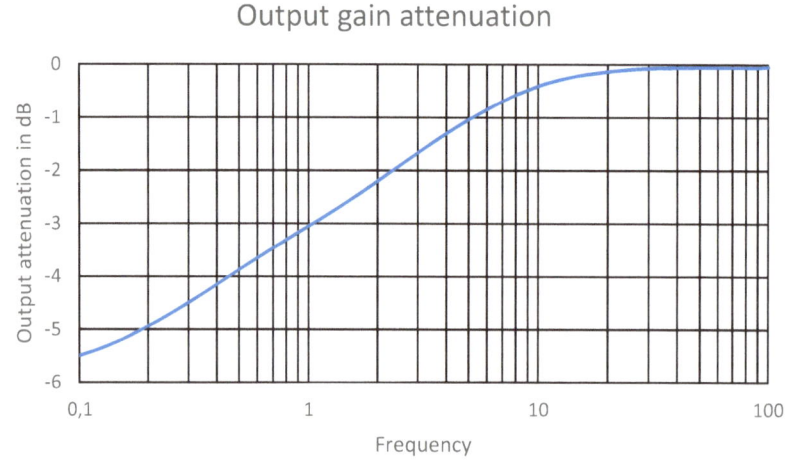

Figure 14: Output gain attenuation with self-bias and bypass capacitor.
Using a capacitor of 150μF we have -3dB output gain attenuation at 1 Hz and -0.4dB at 10 Hz. Attenuation is practically eliminated at 40 Hz.

Chapter 4:
Integrated push-pull vacuum tube amplifier

An integrated amplifier includes all components needed to produce an amplified signal having enough power to drive a speaker, starting from a signal produced by an input source (CD, DAC, radio, etc.). The basic building blocks of a push-pull integrated vacuum tube amplifier are shown in Figure 15.

The input signal is first received by the *input stage*, that operates a pre-amplification of the signal. Push-pull amplifiers, discussed in Section 4.1.4, require two phase-inverted copies of the same signal to be amplified. The *phase splitter stage*, after the input stage, takes a signal and produces two output signals, one 180° phase inverted with respect to the other. The two phase-inverted signals are passed to the *push-pull stage*, composed of two power amplifiers that amplify the two received signals. The *output stage* also includes an output transformer, where the two phase-inverted amplified signals are combined and adapted to be given to the output speaker. The output signal can also be used as a *global negative feedback*, to reduce distortion and noise, at the expense of a reduced gain of the amplifier.

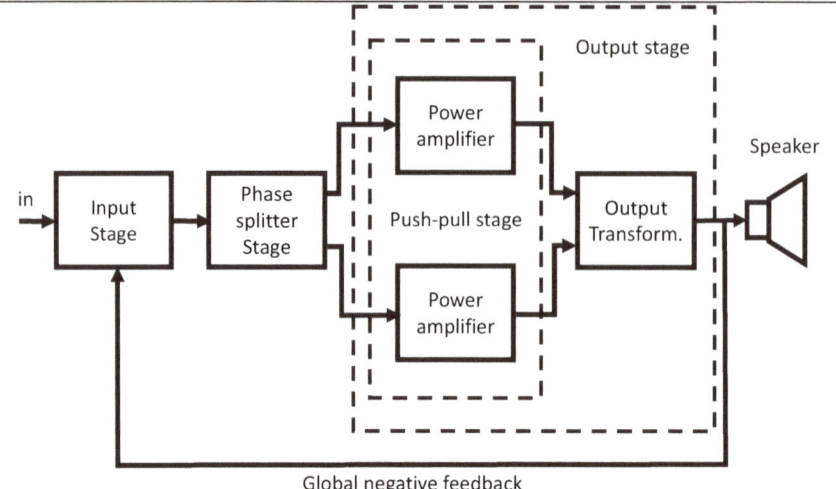

Figure 15: Integrated push-pull vacuum tube amplifier.
The input signal is first handled by the input stage, which pre-amplifies the signal. The signal then goes to the phase splitter stage, which produces two phase-inverted copies of the same signal. These are passed to the two power amplifiers that compose the push-pull stage. The output amplified signal is fed to the output transformer, then to the output speaker. Output signal can also be fed back to the input stage to reduce distortion and noise, at the expense of a reduced gain.

Next sections discuss in more details all these components. We will proceed backward from the output stage to the input stage.

4.1 Output stage or power stage

The task of the *output stage* (or *power stage*) is to amplify the signal, produced by the preceding stage, to have the necessary power to obtain sound from a speaker. We will consider a power amplifier in *Push-Pull* configuration, where a pair of vacuum tubes are used simultaneously to amplify phase inverted copies of the same signal. However, before discussing it, we first need to introduce the *Single Ended* configuration, where a single vacuum tube is used to amplify the signal. The push-pull configuration is then obtained by appropriately pairing two single ended power amplifiers.

4.1.1 Single Ended configuration

Figure 16 depicts the basic schema of Single Ended power output stage built around a tetrode (or a pentode, here the suppressor terminal is not shown) in ultra-linear configuration (see Section 2.2.4) and an output transformer.

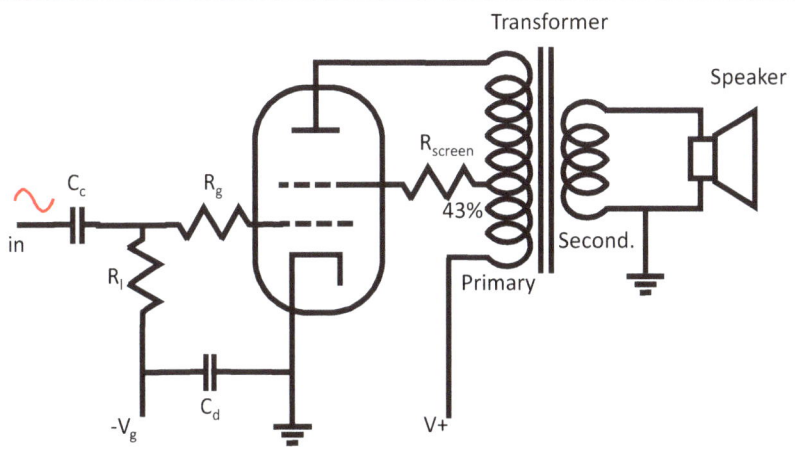

Figure 16: Basic schema of a Single Ended power output stage.
The load of the vacuum tube consists of a transformer that accommodates the amplified signal so that it can be applied to the speaker. The transformer basically transforms the impedance of the speaker into an impedance that can be effectively applied to the anode of the vacuum tube.

The load of the vacuum tube is an output transformer. It is used to couple the vacuum tube and the speaker. In fact, the required load impedance of a power vacuum tube is generally much higher than that provided by commercial speakers. The output transformer transforms the speaker impedance into the load impedance needed by the vacuum tube.

As an example, Figure 17 depicts the average anode characteristics of the EL34 power vacuum tube. Red lines represent various loadlines corresponding to various loads applied to the anode of the vacuum tube. The maximum power that can be dissipated by the anode of an EL34 is 25W, corresponding to the dashed line marked as "$W_a=25W$". All loadlines lay below this dashed line. In the figure, each loadline is labelled with the corresponding load impedance.

We can see that the various possible load impedances vary from 1.1K Ohms to 5.4K Ohms. These values are all far from reasonable values of impedance of commercial speakers, where impedance generally ranges between 4 Ohms and 16 Ohms. Accordingly, the main purpose of the output transformer is to adapt the impedance of the speakers to the impedance required by the power vacuum tubes, as discussed more in details in Section 4.1.2.

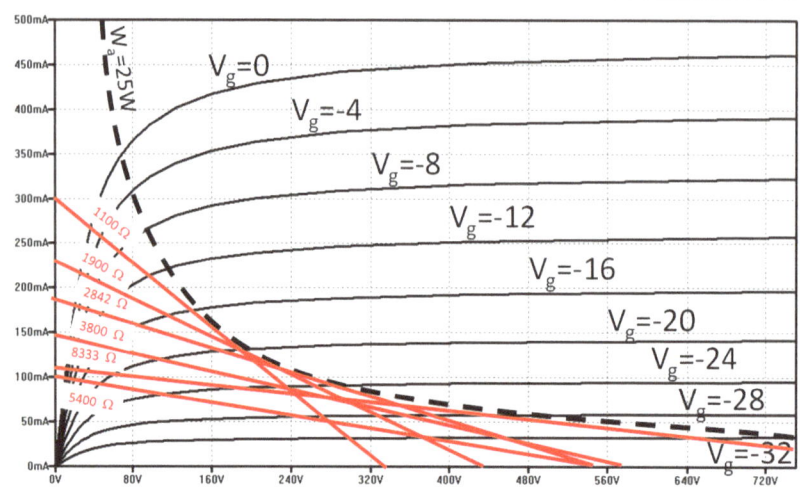

Figure 17: Examples of various possible loadlines, for a power amplifier.
Red lines show various options of loadlines possible with this vacuum tube. In this example possible values range from 1.1K Ohms, to 5.4K Ohms.

The DC voltage V+ is applied to the anode of the power vacuum tube through the centre tap of the output transformer primary. The output transformer also eliminates the DC applied to the anode, so that speakers just see the AC corresponding to the amplified signal.

The screen terminal of the power vacuum tube is connected to a tap from the primary of the output transformer, corresponding to the wanted percentage of the anode output signal, to obtain the ultra-linear configuration (see Section 2.2.4). Many power amplifier designs use a percentage around 43% of the signal to be given to the screen.

The resistor R_{screen}, called the *screen stopper*, which connects the screen to the transformer screen tap, is mainly used to limit the current from the screen, and to avoid parasitic oscillations of the circuit. Screen current becomes dangerous when the power vacuum tube is overdriven. Overdrive should be avoided in Hi-Fi amplifiers, even if it is generally used in guitar amplifiers. Typically, values around 1K Ohms 2W are used. Smaller values or no resistance at all, is sometimes used in Hi-Fi amplifiers.

The resistor R_g, called the *grid stopper*, is used to block high frequency parasitic oscillations and reduce radio interference. The internal components of a vacuum tube produce some parasitic capacitances, generally referred as the *Miller effect*. The grid stopper forms a low-pass filter with these capacitances. The value of the grid stopper mainly depends on the capacitances. For example, values around 4.7K Ohms are generally used for EL34 power vacuum tubes.

R_l is the *grid leak resistor*. The vacuum tube receives the grid bias voltage through this resistor. R_l offers a high impedance path to ground to the AC signal coming from the previous stage and, as we will see in section 4.2.1, contributes to determine the loadline of the previous stage. Finally, R_l has also the important function of discharging the positive charge that might accumulate on the grid because of the gas ions forming in the vacuum tubes. This effect is very dangerous since it can cause *thermal runaway* and destroy the vacuum tube. When positive charges accumulate on the grid, the grid becomes less negative and the vacuum tube conducts more. More current in the vacuum tube produces more gas ions, which accumulate on the grid, and more current that will damage the vacuum tube itself. The correct value for the grid leak is a matter of compromise. Large values are preferable to provide previous stage output signal with a high impedance to ground. Small values are preferable to better help the grid to discharge the accumulated positive gas ions. Datasheets generally provide maximum allowed values for the grid leak, in terms of maximum impedance from the grid to the cathode. For instance, Philips EL34 datasheet specifies maximum 0.5M Ohms in class B, or 0.7M Ohms in class A or AB.

When fixed biasing is used, as in our example, a high value *decoupling capacitor* C_d is also generally connected, from the $-V_g$ side of R_l, to ground. This capacitor, discussed in Section 3.6.1, prevents the input signal from reaching the grid of other vacuum tubes, by providing these residual signals with a very low impedance path to ground.

The capacitor C_c is the *coupling capacitor*. It blocks DC in the signal arriving from previous stage. AC continues to the grid through the grid resistor, and to ground through the grid leak resistor. The coupling capacitor C_c and the grid leak R_l form together a high-pass filter. The value of the capacitor has to be chosen according to the desired low cut-off frequency. Note that no current goes in the vacuum tube, since the grid has a very high impedance and receives just a voltage signal.

> **Example 6: Inter-stage coupling capacitor of the power stage**
>
> Suppose the grid leak is 200K Ohms and we want a cut-off frequency at 7 Hz. We can use the low-pass filter equation to obtain the capacitance of the coupling capacitor:
>
> $$C_c = \frac{1}{2\pi \cdot R_l \cdot f} = \frac{1}{2\pi \cdot 200 \text{kohm} \cdot 7 \text{Hz}} = 0.11 \mu F$$

4.1.2 Impedance of an output transformer

In an output transformer, with one primary end connected to the anode of a vacuum tube, the ratio between the number of turns, of the primary and the secondary, determines the AC impedance seen by the anode, when a load (a speaker in our case) is connected to the secondary.

Let n_p and n_s be respectively the number of turns of the primary and secondary. Let Z_p, and Z_s, be respectively the impedance seen at the primary and the impedance applied to the secondary (that is the impedance of the speaker). We have that the ratio between the primary and secondary impedances is equal to the square of the ratio between the primary and secondary turns. More formally:

$$\frac{Z_p}{Z_s} = \left(\frac{n_p}{n_s}\right)^2.$$

It is worth noting that, according to this equation, we can change the impedance, seen by the anode of the vacuum tube, which is the impedance of the primary of the transformer, by changing the impedance of the speaker. The impedance of the speaker is reflected to the primary of the transformer, according to the square of the ratio between primary and secondary turns.

Example 7: Impedance of an output transformer

Suppose that we use a speaker with an impedance of 8 Ohms and we want to have a load of 3.8K Ohms to be seen by the anode. Then, we need a transformer where the square of the ratio between the primary and secondary turns is 3.8k/8=475. If we connect a speaker of 4 Ohms to the secondary of the same transformer, then the impedance seen by the anode will be 475·4 Ohms=1.9K Ohms. If we connect a speaker of 16 Ohms, we will have 475·16 Ohms=7600 Ohms.

It is also worth mentioning that speakers, generally do not have a constant impedance. Rather, the speaker's impedance varies with the frequency of the signal being reproduced. Figure 18 shows the real impedance of speakers rated for 11 Ohms.

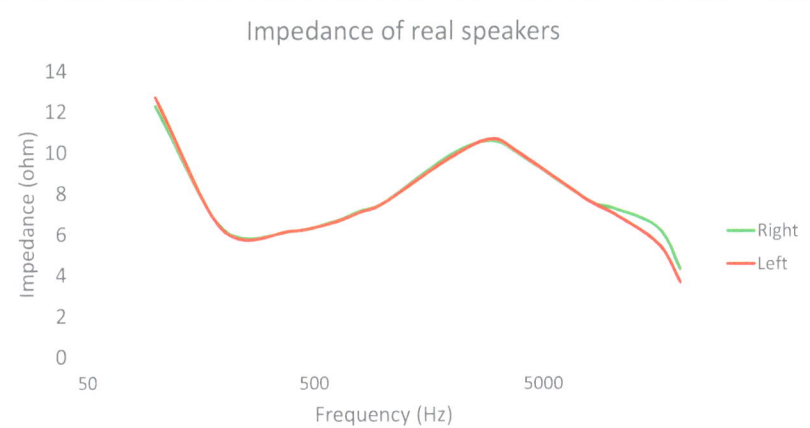

Figure 18: Impedance of real speakers.
In a real speaker the impedance varies with the frequency of the sound being reproduced. The graph above refers to real impedance measured on speakers rated for 11 Ohms. Real impedance varies from 4 Ohms at frequencies around 20K Hz, to 11 Ohms around 3K Hz. There is also small difference between right and left speaker.

In this example, the impedance is very high around 100 Hz. It goes to 6 Ohms at 400 Hz, then jumps to 11 Ohms at 3K Hz, and then back at 4 Ohms to 20K Hz. This means that the slope of the loadline might vary significantly depending on the reproduced frequencies. The loadline, which is typically plotted as single line, is in fact a blurred area around the thin line.

4.1.3 Reactive load and loadline computation

A transformer is a reactive load that offers an impedance just when an AC signal goes through its primary. The transformer primary, practically, does not offer impedance when just DC is applied to it. When no signal is applied to the grid, the vacuum tube is in a quiescent state and no AC signal is produced at its anode. In this case, only the DC current goes through the transformer primary and no impedance is seen by the anode. In addition, no signal is transferred from the primary to the secondary of the transformer.

Since there is not impedance and there is no voltage drop, the anode receives the full V+ voltage, at the quiescent reactive operating point. Accordingly, the quiescent current is the one associated with V+ along the plot corresponding to the chosen grid bias voltage.

When the anode produces an AC signal, then the transformer offers a resistance and the anode voltage and current start oscillating around the operating point along the *reactive loadline*. The reactive loadline is parallel to the *resistive loadline*, which can be computed as described in Section 3.2, and shifted so that it goes through the reactive operating point. Figure 19 shows the resistive loadline (green line) and the reactive loadline (red line) in correspondence of a load of 3.8K Ohms, a voltage V+=400V, and a bias current of 40 mA. The red line is parallel to the green line and shifted higher so that it passes through the reactive operating point (red spot).

Example 8 clarifies this.

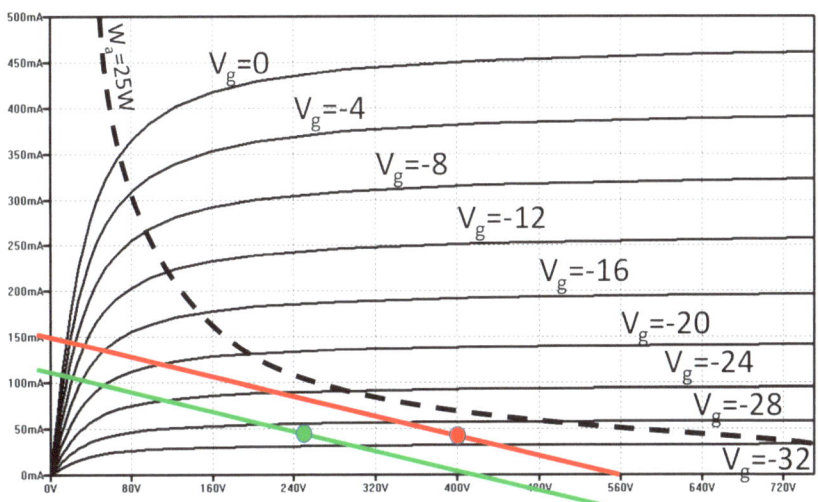

Figure 19: Reactive loadline.
In case of a resistive load, with an anode voltage of 400V, the loadline would be along the green line, and the operating point with a bias current of 40mA would be at the green spot. However, an output transformer has a reactive behaviour. In this case, there is a resistance just when an AC signal goes through it. There is no resistance when the vacuum tube is quiescent, so the quiescent voltage remains 400V independently of the bias current. In case of a bias current of 40mA the operating point is depicted by the red spot. When the vacuum tube amplifies an AC signal, an AC current goes from anode to cathode and trough the transformer, which now offers a resistance. The loadline, in this case, as depicted by the red line, is parallel to the resistive loadline, and shifted so that it passes through the reactive operating point. Note that the voltage reached by the anode, in case of a reactive load, might be higher than the voltage V+ applied to the transformer primary. This is due to the transformer reacting to current variations, virtually accumulating and releasing energy accordingly.

Example 8: Determining the reactive loadline

If we had a resistive load, we could have computed the loadline as we described in Section 3.2. For instance, suppose we had a load of 3.8K Ohms and a voltage V+ of 400V. At no conduction, the anode voltage would have been 400V. At full conduction, the anode voltage would have been 0 and the current 400V/3.8K Ohms=105mA. In this case, the resistive loadline would have been represented by the green line in Figure 19. A bias current of 40mA would have set the quiescent operating point at the green spot in the figure.

However, in our case the load is an output transformer, which has a reactive load. The transformer offers resistance just to AC signals that go through it. There is almost no resistance to DC. More specifically, when the vacuum tube is in a quiescent state, just DC goes from the anode to the cathode, corresponding to the bias current. In this case, the transformer does not offer resistance. This means that the voltage applied to the anode is the same than the V+ voltage applied to the transformer primary, 400V in our example. Therefore, a bias current of 40mA sets the *reactive operating point* as depicted by the red spot in Figure 19. When an AC signal goes through the vacuum tube, the transformer offers its 3.8K Ohms of impedance, so the anode voltage and current start oscillating along the *reactive loadline* represented by the red line in the figure. Note that reactive loadline is parallel to the resistive loadline and shifted higher so that it goes through the reactive operating point. It might seem strange that the anode voltage can now reach values higher than the 400V applied to the transformer primary. However, this is due to the transformer reacting to current variations, virtually accumulating and releasing energy in accordance to these variations.

4.1.4 Push-Pull configuration

In a push-pull amplifier, two power vacuum tubes are used to amplify phase-inverted copies of the same signal. The phase-splitter stage, (discussed later in Section 4.2), before the push-pull power stage, creates the two phase-inverted copies of the same input signal. The two phase-inverted signals, amplified by the two power tubes, are combined together using an output transformer appositively conceived for push-pull configuration.

The basic schema of a push-pull amplifier stage is depicted in Figure 20. It is composed of two vacuum tubes having an identical configuration and circuitry. The primary of the push-pull output transformer has a centre tap that receives the V+ voltage. Anodes of the two vacuum tubes are connected to the two ends of the output transformer primary. Similarly, the two taps for ultra-linear configuration are connected to the two screens of the vacuum tubes. If we consider just half transformer and one vacuum tube, the schema is practically identical to the single ended configuration. A push-pull stage can be seen as two single ended stages connected together through the push-pull output transformer. When no signal is applied to phase splitter, the two power vacuum tubes receive no signal and just the bias current goes from the anodes to the cathodes. The current enters from the transformer primary centre tap and flows in opposite direction through the transformer toward the first and the second vacuum tube.

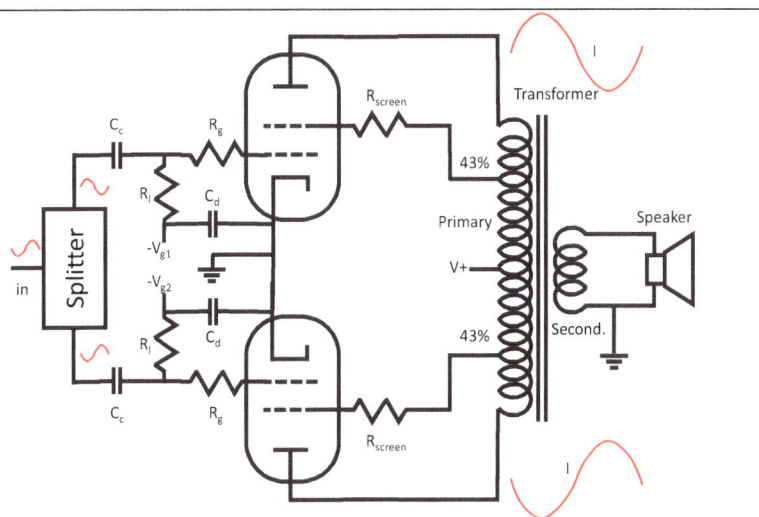

Figure 20: Basic schema of a push-pull amplifier.
The phase splitter, which will be discussed later, takes the input signal and returns two copies of the same signal each with phase inverted with respect to the other. The two signals are given to two vacuum tubes that amplify them separately. The primary of the output transformer, used for a push-pull stage, has a centre tap that takes the high voltage. The two ends of the primary go to the anodes of the two vacuum tubes. Given that the two vacuum tubes amplify signals that are 180° one from the other, in the two ends of the transformer current varies symmetrically. That is, when in one end current increases, in the other end current decreases.

When a signal is received by the two vacuum tubes, the two amplified signals are identical, with inverted phases, and make the current to vary symmetrically, with respect to the bias current, in the two halves of the push-pull transformer. When in one half of the transformer current increases, with respect to the bias current, in the other half current decreases. The same effect happens to the voltage measured at the two anodes of the two vacuum tubes. On the load line, this effect is seen as movements in opposite directions, as depicted in Figure 21. While one tube moves following the green arrow, the other one moves following the yellow arrow, and vice versa.

The power vacuum tubes used in a push-pull power stage should be perfectly matched to guarantee a symmetrical behaviour. In order to fine tune the two power vacuum tubes, the two bias voltages -V_{g1} and -V_{g2} should be adjustable so that the quiescent bias current is exactly the same in the two tubes.

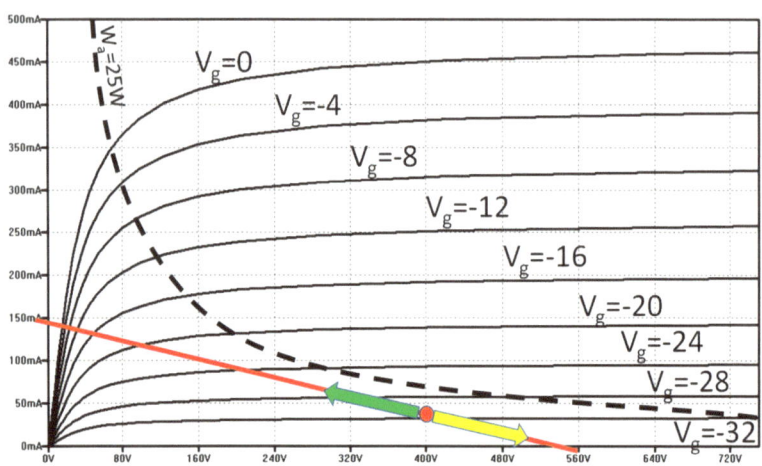

Figure 21: Dynamic behaviour of a push-pull amplifier.
When quiescent, the two power vacuum tubes are at the same operating point, depicted by the red spot. When a signal is amplified, since the vacuum tubes receive signals with inverted phases, they move in opposite directions. When one conducts more, as depicted by the green arrow, the other conducts less, as shown by the yellow arrow.

4.1.5 Push-pull advantages

Push pull amplifiers offer some advantages with respect to the single ended ones. Push-pull transformers core can be made smaller than single ended ones and without air gaps. In fact, the current from the centre tap flows in opposite directions through the two half of the transformer and one current cancel the electro-magnetic effect of the other. Two additional advantages are the reduction of the harmonic distortions, and the increased output power, as discussed below.

When the vacuum tube's operating range is situated in an asymmetrically nonlinear area, that is when the grid voltage lines intersect more densely the loadline on one side rather than the other, second order harmonics are generated. For instance, in Figure 21, grid voltage lines intersect the loadline more densely with an anode voltage higher than 400V, than with an anode voltage lower than 200V. Therefore, half signal is amplified very differently than the other half signal. Remember that the input signal corresponds to different grid voltages, that is different points, identified by the intersection of the grid voltage lines with the loadline. Figure 22 shows the effect of this nonlinearity. The upper plot shows the anode current of two vacuum tubes in push-pull. Since each vacuum tube amplifies a phase inverted signal, with respect to the other, the currents at the anodes of the two vacuum tubes are phase inverted as well. Note that the current signal is somehow flattened in the lower parts of the plots. This is because, as can be seen in Figure 21, there is less current variation on the right side of the loadline, compared to the left side of the loadline, in correspondence of the same grid voltage variation. This asymmetric behaviour introduces even order (mainly second order) harmonic distortions. However, given that the two signals are phase inverted, signal flattening occurs in one vacuum tube at time. When one vacuum tube is operating in a denser area, the other is operating in a less dense area. When one vacuum tube flattens the signal, the other vacuum tube does not. The two phase inverted signals are combined in the output transformer and even order harmonics, generated by the power vacuum tubes, are significantly attenuated, or cancelled, as can be seen in the lower plot of the figure. Note, however, that harmonic distortions introduced by previous stages or third order harmonics, introduced by symmetric nonlinearity, are amplified and not attenuated. Harmonic distortions produced by stages preceding the push-pull stage can be attenuated using global negative feedback, as will be discussed later in Section 4.4.

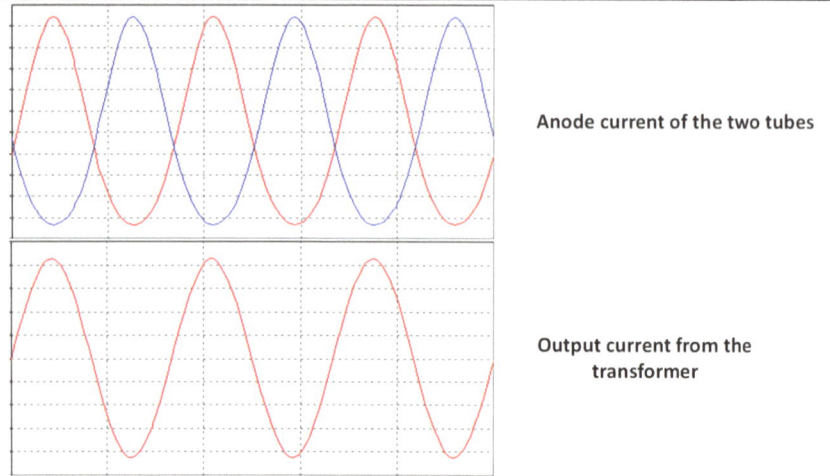

Figure 22: Even order harmonic distortion in a push-pull amplifier.
The upper plot shows the anode current of two vacuum tubes configured as push-pull when they amplify a signal. Note that when the amplified signal produced by one vacuum tube flattens in the lower part, due to the introduction of even order harmonics, the other vacuum tube compensates. The result is that the signal coming out from the output transformer, shown in the lower plot, is again an almost perfect sinusoidal signal, where second order harmonics are significantly reduced.

The push-pull configuration also offers the possibility of increasing the headroom, that is the operating range along the loadline. Larger headroom implies higher output power. When the amplitude of the input signal is too large, either the vacuum tube saturates or cuts-off (see Section 3.3). In a Single ended configuration, this is avoided by setting the bias so that it works in Class A condition (see Section 3.5) and taking care that the input signal is not too large, limiting therefore the amplifier headroom. However, in a push-pull amplifier we can set the bias so that vacuum tubes operate in class AB, with very low harmonic distortions. In class AB, one vacuum tube amplifies most of the signal and cuts-off when the input signal is below a certain threshold. Cut-off happens when the signal reaches the point where the loadline intersects the horizontal axis. When operating in class AB, given that the two vacuum tubes amplify phase inverted signals, if one vacuum tube cuts-off, the other vacuum tube still amplifies the inverted signal. This is depicted in the upper plot of Figure 23. The output transformer combines the two signals and an almost perfect signal is sent to the speaker, as shown in the lower plot of

Figure 23. Operating in class AB allows setting the bias so that the vacuum tubes cuts-off alternatively, increasing the overall headroom.

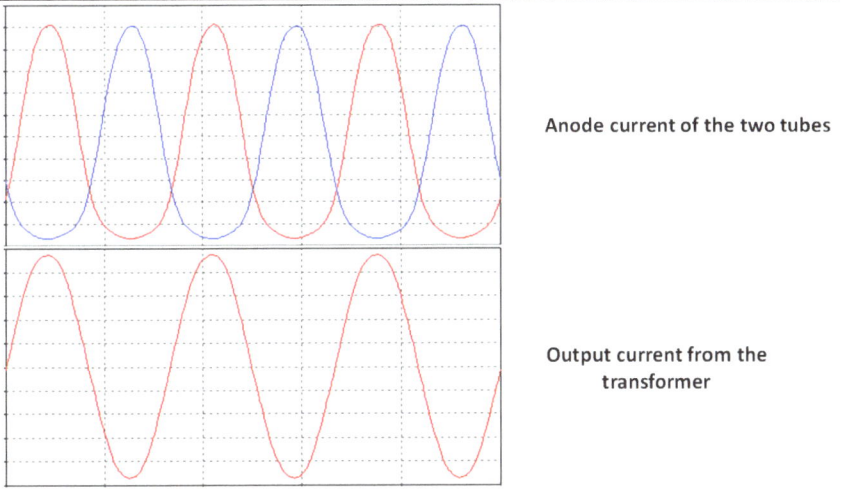

Anode current of the two tubes

Output current from the transformer

Figure 23: Push-pull amplifier operating in class AB.
When operating in class AB, one vacuum tube amplifies just part of the whole signal, as shown in the upper plot. However, when one vacuum tube cuts-off the other still amplifies the signal, and vice versa. When the two signals are combined, in the output transformer, the original input signal is practically restored as shown in the lower plot.

4.1.6 Push-pull loadline in class AB

In a push pull output transformer, the impedance rating is referred to the whole impedance seen from one end to the other end of the primary. However, two vacuum tubes simultaneously use the transformer and each tube sees half impedance of the transformer. Therefore, when using a push pull transformer with an impedance of 7.6K Ohms, each anode sees an impedance of 3.8K Ohms. The loadline, for the power vacuum tubes of a push-pull amplifier, has to be drawn considering half the transformer impedance. The method for drawing the loadline, along with the relationships between the actual transformer impedance and the speaker impedance, has already been discussed in Section 4.1.2.

However, at a closer look, in case of a push-pull power stage, we have an additional consideration to make. We just said that the impedance seen by each anode is half that of the end-to-end impedance of the transformer primary. This is true only when both vacuum tubes are conducting simultaneously, for instance, when both vacuum tubes operate in class A. However, class AB operation is generally used In a push-pull power stage. In a class AB amplifier, as discussed in Sections 3.5 and 4.1.5, one vacuum tube might quit conducting while the other is fully active, and vice versa. When one of the two vacuum tubes does not conduct, just half transformer is used. In this case, the impedance seen by the anode is not half than that of the entire transformer. In fact, we already said that the impedance of the transformer goes with the square of the ratio between the number of turns, of the primary and the secondary of the transformer. When we use half transformer, we use half turns of the transformer primary and the impedance seen at the anode is one fourth of the entire transformer impedance. This can be verified using the equations discussed in Section 4.1.2.

Figure 24 clarifies these aspects. When the two vacuum tubes amplify each, a signal inverted with respect to the other, their operating points move in opposite directions along the loadline, starting from the quiescent operating point. The green and the yellow arrow represent the simultaneous position of the two vacuum tubes along the loadline. Suppose the signal amplified by the yellow vacuum tube is such that it reaches the bottom of the graph and quits conducting. The green vacuum tube, at this point, sees a different impedance at its anode and the loadline slope changes accordingly. From this point on, the green arrow follows the solid red line, rather than the dashed line corresponding to the continuation of the previous loadline. When the signal being amplified is such that the yellow vacuum tube starts conducting again, the green arrow will be back at the position where the impedance first changed. At this point, the green vacuum tube sees the original impedance again, and continues along the solid red line. Note, of course, that the impedance change is not abrupt as depicted in the figure. When the yellow vacuum tube is in the process of quitting conduction, the impedance seen by the other tube changes gradually.

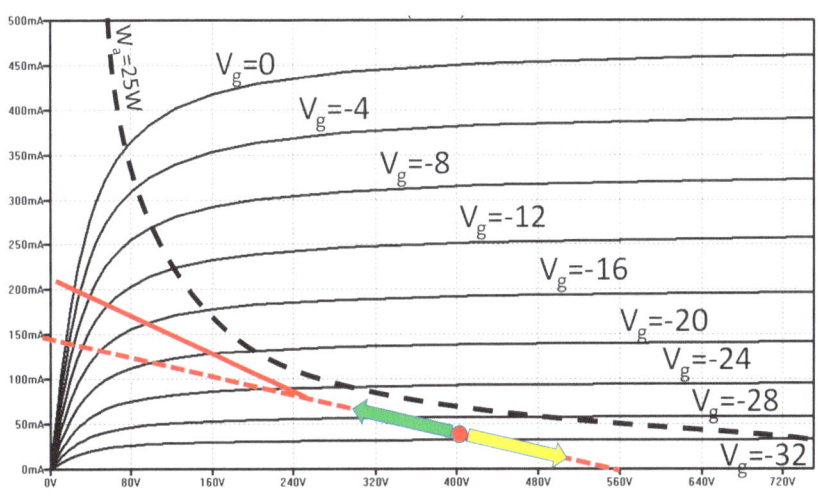

Figure 24: Loadline in a class AB push-pull amplifier.
In a push-pull transformer, operating in class AB, the loadline is not a straight line. While both vacuum tubes are conducting, the impedance seen at the anode is half than that of the entire transformer. When one of the two vacuum tubes quits conducting, just half transformer is used, and given that the impedance goes with the square of the turn ratio, the impedance will be one fourth of the entire transformer impedance. In the picture, when the yellow arrow, representing one power vacuum tube, reaches the bottom, the green arrow, representing the other vacuum tube, continues following the solid red line, rather than the dashed line.

4.2 Phase splitter stage

The purpose of the phase splitter stage is to produce two identical phase inverted output signals from the input signal. The two phase inverted signals are then passed to the push-pull power stage.

The most common schemas are the *paraphase* phase splitter, the *long-tailed pair* phase splitter, and the *concertina* phase splitter. For simplicity, here we will discuss the concertina phase splitter only. Despite its simplicity, a well-designed concertina phase splitter stage offers very high quality. Its only drawbacks might be that, as we will see later, it does not produce any gain on the input signal.

4.2.1 Concertina phase splitter

The basic schema of a *concertina*, or *cathodyne*, phase splitter is given in Figure 25. Two resistors of exactly the same resistance $R_a=R_k$ are connected, respectively, from the high-tension V+ to the anode, and from the cathode to the ground. In this way, the two resistors, along with the vacuum tube internal resistance, form a voltage divider. When grid voltage increases, the resistance offered by the vacuum tube decreases. When the vacuum tube resistance decreases, the voltage at the anode decrease and the voltage at the cathode increases of exactly the same amount, provided the two resistors are perfectly matched. Similarly, when the grid voltage decreases, the anode and the cathode voltages, respectively increase and decrease of exactly the same amount. Consequently, two phase-inverted signals of exactly the same amplitude are taken at the anode and the cathode of the vacuum tube. These two signals are passed to the push-pull vacuum tubes of the next stage.

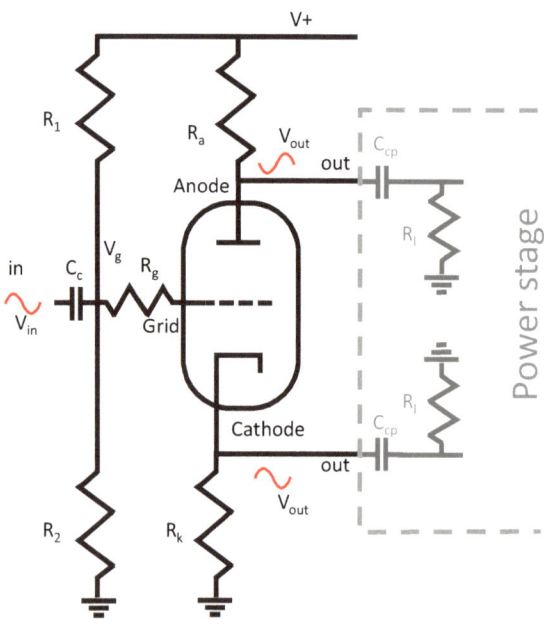

Figure 25: Basic schema of a Concertina phase splitter.
The concertina phase splitter uses two resistors of the same resistance $R_k=R_a$ connected to the anode and to the cathode of a vacuum tube. Voltage variations of at the anode and at the cathode of the vacuum tube, due to a signal applied to the grid, are of the same amplitude and have inverted phases. A voltage divider, built with appropriate values of R_1 and R_2, is used to provide the gird with the correct bias voltage. The coupling capacitor C_c is used to isolate the previous stage from the bias voltage applied to the grid.

The grid bias voltage should be, as usual, negative with respect to the cathode voltage. In this configuration, the cathode voltage is higher than ground due to the effect of the cathode resistor R_k. Therefore, the voltage

V_g should be lower, than the cathode voltage, of an amount equal to the wanted bias voltage. Bias voltage to the vacuum tube grid can be obtained using a voltage divider composed of the resistors R_1 and R_2.

> **Example 9: Biasing the Concertina phase splitter**
>
> Let I_b be the wanted bias current. The quiescent cathode voltage is $V_k = R_k \times I_b$.
>
> Suppose we want a bias current I_b of 1 mA, corresponding to the red spot in Figure 26, and that the cathode resistance R_k is 100K Ohms. The cathode voltage is V_k=100K Ohms × 1 mA= 100V. Given that R_a is equal to R_k, the voltage drop, produced by the anode resistor, is also 100V and the anode voltage is V_a=380V-100V=280V. Consequently, the anode to cathode voltage is V_a- V_k=280V-100V=180V.
>
> According to Figure 26, the bias voltage applied to the grid should be approximatively 1.4V below the cathode voltage. This corresponds to 100V-1.4V=98.6V from ground.
>
> To compute the correct values for R_1 and R_2 we can use the voltage divider equation:
>
> $$V_g = \frac{V_+ \cdot R_2}{(R_1 + R_2)}, \text{or} \frac{R_1}{R_2} = \frac{V_+}{V_g} - 1$$
>
> Suppose high-tension V+ is 380V, the ratio between the two resistors must be (380V/98,6V)-1≈ 2,85. Here large values have to be used, to avoid dissipating too much power. For example, R_1=1,2M Ohms and R_2=560K Ohms are values that give a ratio very close to what we need.

The grid stopper resistor R_g has the purpose of blocking very high frequencies and parasitic oscillations, by creating a low-pass filter together with the internal vacuum tube capacitance, due to the Miller effect, as already discussed in Section 4.1.1. The grid stopper resistor must be soldered directly on the pin of the vacuum tube socket. For instance, values around 47K Ohms are generally used for vacuum tubes like the 12AX7.

The coupling capacitor C_c isolates previous stage from the bias voltage applied to the grid. The coupling capacitor together with R_1 and R_2, seen as parallel resistors when handling an AC signal, form a high-pass filter. C_c has to be chosen according to the wanted low cut-off frequency.

Example 10: Inter-stage coupling capacitor in a concertina phase splitter

Suppose R_1=1,2M Ohms, R_2=560K Ohms, and we want a cut-off frequency at 7 Hz, given that R_1 and R_2 in parallel are R=382K Ohms, we have:

$$C_c = \frac{1}{2\pi \cdot R \cdot f} = \frac{1}{2\pi \cdot 382k\,ohm \cdot 7Hz} \approx 0.06 \mu F$$

4.2.2 DC and AC loadline of a concertina

The values of the anode and cathode resistors determine the load seen by the vacuum tube, and consequently the loadline. The total load seen by the vacuum tube at the quiescent state, that is when there is only DC bias current traversing the vacuum tube, is R_a+R_k. However, when an AC signal is applied to the grid, the two output signals also traverse the coupling capacitors C_{cp}, placed between the concertina and the power stage, and go to the two grid leak resistors R_l of the power stage itself. The two grid leak resistors are in parallel respectively to the anode and cathode resistors of the concertina. The total load, in this case, is the sum of the resistances resulting from the parallel effect of cathode resistor and one grid leak, on one side, and the anode resistor and the other grid leak, on the other side. In a few words, as shown in Figure 26, we have two different loadlines. One loadline, which we call the *DC loadline*, has to be considered when the vacuum tube is in a quiescent state, to analyse the load and the grid bias needed for an optimal quiescent operating point. The other loadline, which we call the *AC loadline*, has to be taken into account to analyse what happens when a signal goes through the vacuum tube and traverse the coupling capacitors. Note that, since the signal oscillates around the quiescent operating point, the AC loadline intersects the DC loadline exactly at this point, identified by the red spot in the figure.

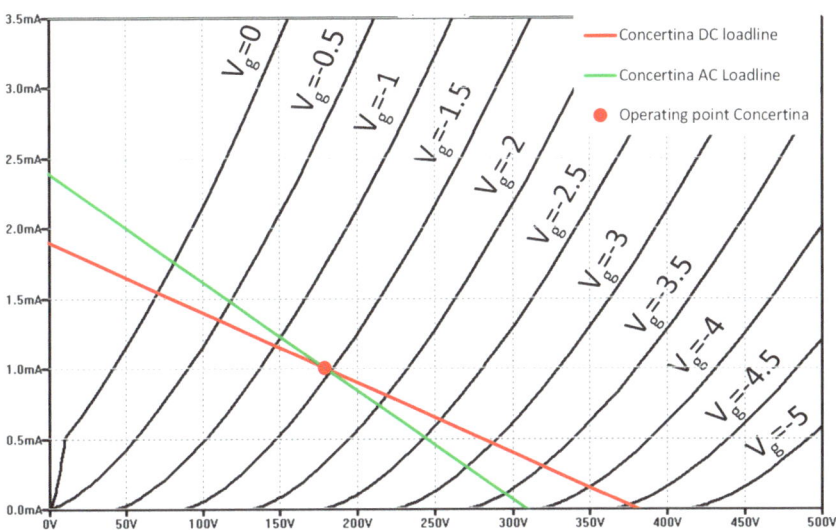

Figure 26: DC and AC loadline of a concertina phase splitter.
The concertina phase splitter has two loadlines. The DC loadline has to be considered when the vacuum tube is in a quiescent state. In this case, just the anode and cathode resistors contribute to the overall load seen by the vacuum tube. When a signal is applied to the grid, the signals at the anode and cathode of the vacuum tube go through the coupling capacitors C_{cp} and from these to the grid leak resistors of the power stage, which are now parallel to the anode and cathode resistors. The total load seen by the vacuum tube is reduced and the AC loadline has to be considered in this case.

Example 11: DC and AC loadlines in a concertina phase splitter

Suppose that $R_a = R_k = 100K$ Ohms, that the two grid leaks of the power stage are $R_l = 180K$ Ohms, and that the high-tension is V+=380V.

The total DC load is 200K Ohms and produces the DC loadline depicted by the red line in Figure 26. The AC load seen, respectively at the anode and at the cathode, is the result of 100K Ohms in parallel with 180K Ohms, which is approximately 64K Ohms. The total AC load in this case is 128K Ohms, which gives the green AC loadline in Figure 26.

4.2.3 Gain of the concertina phase splitter

The cathode resistor R_k puts the vacuum tube of the concertina phase splitter under heavy local negative feedback. As explained in Section 3.6.3, the cathode resistor tends to vary the cathode voltage in the same direction than the grid voltage variation, thus opposing the grid to cathode voltage variation and reducing the gain of the voltage amplifier. Given $R_a=R_k$, if $\mu \cdot R_k \gg r_a$, and $\mu \gg 2$, the gain of this configuration is

$$A^{conc} = \mu \cdot \frac{R_k}{r_a + R_k + R_k(1+\mu)} \approx \frac{\mu}{2+\mu} \approx 1.$$

For instance, using a 12AX7 vacuum tube, with $R_a=R_k>$100K Ohms, since the amplification factor is $\mu=100$, the gain is around 0.98, at both outputs. In practice, in this case, the concertina simply takes the input signal and produces two outputs signals (one inverted with respect to the other) practically of the same amplitude of the input signal.

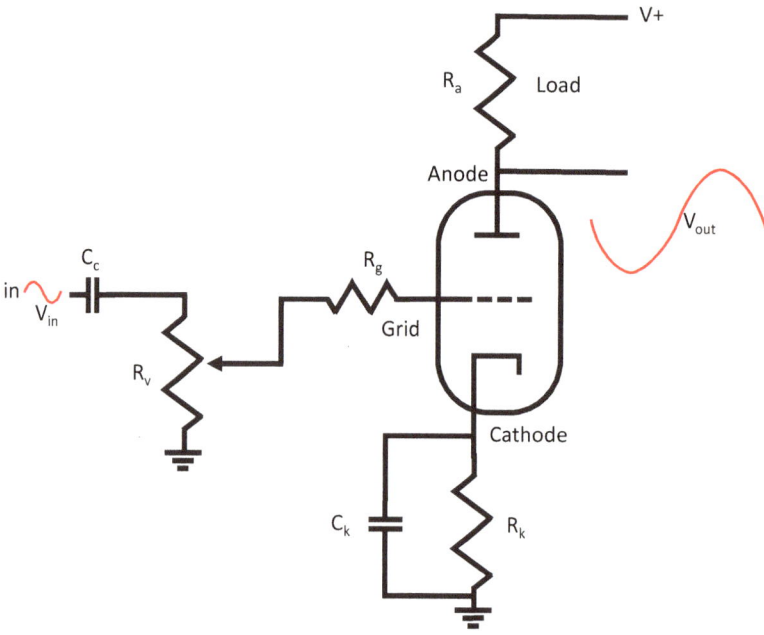

Figure 27: Basic schema of the input stage.
The input stage is basically a voltage amplifier. It takes the input signal, coming from an external source, and applies it to the grid of the vacuum tube. Cathode bias is generally used in this stage, and the cathode resistor is generally bypassed by a capacitor to reduce the local negative feedback and increase the gain.

4.3 Input stage

The input stage is basically a voltage amplifier, which was already discussed in Chapter 3:. An example of input stage is shown in Figure 27.

A resistive load R_a is connected to the anode of the vacuum tube. The output voltage is taken at the anode itself, before the load. The grid bias voltage for the input stage is obtained by using self-biasing with a cathode resistor R_k at the vacuum tube cathode. The cathode resistor is bypassed by a capacitor C_k to reduce the local negative feedback, produced by the cathode resistor, and to increase gain. Cathode biasing and local negative feedback details were already discussed in Section 3.6.2.

Next example shows how to draw the loadline, set the quiescent operating point, and chose the cathode resistor for self-biasing the input stage.

Example 12: Loadline and bias of the input stage

Suppose, for instance, the input stage is built using a 12AX7 vacuum tube, the high-tension V+ is 300 V, and the load R_a is 220K Ohms.

When the vacuum does not conduce, the anode to cathode voltage is 300V. In the theoretical case that the vacuum tube does not offer any resistance, the anode current is 300V/220K Ohms=1.35mA. By connecting these two points, we obtain the loadline plotted in Figure 28. A good operating point is identified by the red spot. It corresponds to a bias current I_b of 0.65mA and an anode to cathode voltage of V_a 160V. This can be obtained with a grid bias voltage of -1.5V. Since we are using cathode biasing, the grid is at ground level and we have to elevate the cathode voltage to 1.5V, by computing an appropriate value of the cathode resistor R_k. By using the Ohm's law, we have that $R_k=V_k/I_b$=1.5V/0.65mA=2.3K Ohms. The closest standard resistance is 2.2K Ohms, which is a good approximation.

The bypass capacitor C_k has the purpose, as discussed in Section 3.6.2, to reduce local negative feedback and increase gain. Small capacitance values increase gain just for high frequencies, high ones increase gain also for low frequencies. For instance, in our case, using the calculations discussed in Section 3.6.3, we determine that a value of $100\mu F$ is enough to bypass and increase gain at all audible frequencies.

The grid stopper resistor R_g is used to block very high frequencies that can enter the circuit and parasitic oscillations, by forming a low-pass filter with the internal vacuum tube capacitance. As we already said, values around 47K Ohms are generally used with 12AX7 vacuum tubes.

The use of a grid stopper resistor is particularly important here. We are at the very beginning of the amplifier stages and signals that have to be amplified are very small. All noises, interferences, parasitic oscillations are here significantly amplified through all the remaining stages. For instance, consider that the wire connecting the input jack to the gird cannot be generally very short, for practical assembling issues, and it acts as an antenna capturing electromagnetic interferences, which must be blocked before being amplified.

The potentiometer R_v is used to control the volume of the amplifier, that is the amount of input signal applied to the grid of the input stage vacuum tube. The potentiometer also acts as a grid leak resistor and forms, with the coupling capacitor C_c, a high-pass filter that blocks the unwanted low frequencies. The coupling capacitor C_c also isolates the input stage from possible DC coming from the external input source.

Example 13: Coupling capacitor of the input stage

The value of the coupling capacitor C_c can be computed using the high pass filter formula. Suppose the R_v potentiometer is 100k Ohms, and we want to stop all frequencies below 7 Hz. We have:

$$C_c = \frac{1}{2\pi \cdot 100 kohm \cdot 7Hz} \approx 0{,}22 \mu F .$$

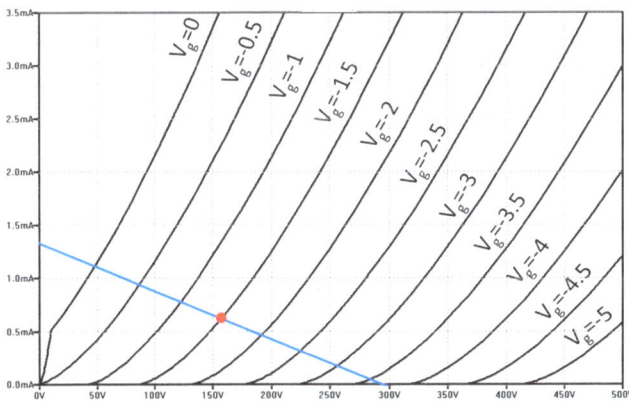

Figure 28: Loadline and operating point of the input stage.
Loadline and operating point obtained with a 12AX7 vacuum tube, with an anode load of 220K Ohms, and a cathode resistance of 2.2K Ohms.

4.3.1 Directly Coupled concertina

In the Example 9 we determined that the quiescent cathode voltage of the concertina was 100V. In order to obtain a bias voltage of -1.4V, with respect to the cathode, we had to bring the grid voltage to 100V-1.4V=98.6V, by using a voltage divider. The coupling capacitor had the main purpose of isolating the grid voltage from the anode voltage of the input stage. In fact, in Example 12 , the quiescent anode voltage was 160V. The coupling capacitor, basically, had the purpose of isolating the 98.6V quiescent grid voltage, of the concertina, from the 160V quiescent anode voltage, of the input stage. In this way, just the AC signal was allowed to go from the anode of the input stage to the grid of the concertina.

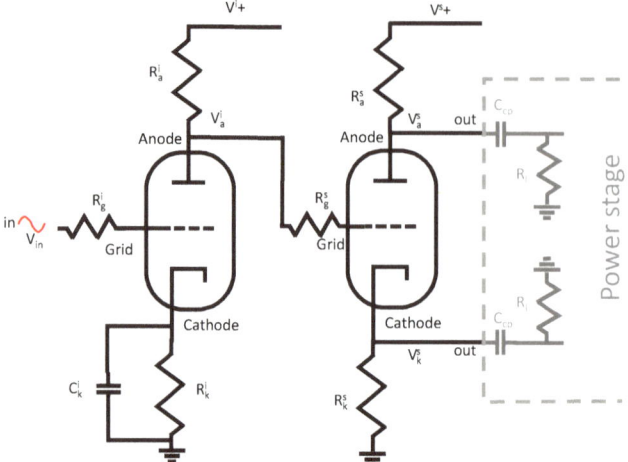

Figure 29: Directly coupled Concertina.
In a directly coupled concertina, bias voltage for the phase splitter grid can be taken directly from the anode of the input stage. In order to do that, loadlines and operating points of both the input and the splitter stages should be chosen so that, in the quiescent state, voltage at the anode should be below the voltage at the cathode of the splitter of a value corresponding to the wanted grid bias voltage. This allows eliminating the coupling capacitor between input stage and phase splitter, and the voltage divider, needed to set the grid bias. Eliminating the coupling capacitor is particularly relevant, since less components along the signal path always improves sonic quality of the amplifier.

However, in many cases, it is possible to set the quiescent operating points of both concertina and input stages so that the quiescent anode to ground voltage of the input stage is exactly what is needed at the concertina grid. This allows directly coupling the input stage and the concertina stage. The input stage quiescent anode voltage is used to bias the concertina, thus eliminating both the coupling capacitor and the voltage divider. Eliminating those components, not only makes the schema simpler and cheaper. It also improves sonic quality of the amplifier. In fact, remember, that all components along the signal path slightly degrade the audio quality of the amplifier. The resulting schema of the Directly Coupled concertina is that shown in Figure 29.

Example 14: Biasing for directly coupled Concertina

Suppose we use the configuration from Example 9 for the concertina stage. We determined that the needed grid to ground voltage is 98.6V, to produce a grid bias voltage (grid to cathode voltage) of -1.4V. Correctly configuring the input stage, and directly connecting the input stage anode to the concertina stage grid, without coupling capacitor, can accomplish to this.

With an input stage high-tension voltage V^i+ of 300V and a load of 220K Ohms we obtain the input stage loadline depicted by the violet line in Figure 30. With a grid bias of -0.7V, we obtain an anode to cathode voltage of 98V and a quiescent current of 0.9mA, as depicted by the blue spot in the figure. The needed grid bias of -0.7V can be obtained, using the cathode biasing technique (see Section 3.6.2). Choosing a cathode resistor 820 Ohms we obtain a cathode elevation of 820 Ohms·0.9mA=0.73V, which is close to what needed. The anode to ground voltage of the input stage is 98V+0.73V=98.73V, which is also very close to the voltage that we need at the grid of the concertina stage.

Using these values, we are able to directly couple input stage and concertina splitter stage, saving components, costs, and improving sonic quality of the amplifier.

Remember that we have also to consider the AC loadline, to check that the concertina operates linearly when handling a signal. Suppose that the power stage grid leak resistors values are 180K Ohms. These resistors, parallel to the anode and cathode resistors of the concertina, give approximately 65K Ohms and produce the AC loadline depicted by the green line in Figure 30, which shows that the concertina operates in a fairly linear area.

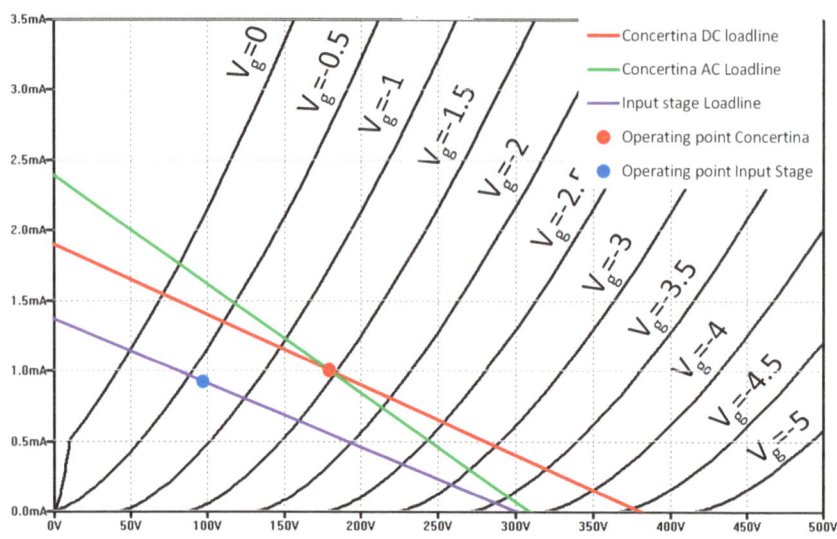

Figure 30: Loadlines and operating points for directly coupled Concertina.
Suppose we set the high-tension voltage of the concertina to 380V, and the anode and cathode resistors both to 100K Ohms. We obtain the DC and AC loadlines depicted by the red and green lines in the graph. Suppose we set the quiescent operating point to 1mA current. This can be obtained with a grid bias voltage of -1.4V. With 1mA quiescent current, the cathode voltage of the concertina is 100V, so the grid voltage should be 100V-1.4V=98.6V. In order to direct couple the input stage and the concertina, the input stage anode voltage should be set to this value. This can be obtained by setting the input stage high-tension voltage to 300V, the load to 220k, and the grid bias to 0.7V. In this way, we obtain the loadline depicted by the violet line, and the wanted quiescent anode voltage, as depicted by the blue spot.

4.4 Global Negative Feedback

Push-pull stages are able to reduce harmonic distortions, specifically second order distortions, produced in the stage itself. The input stage has a single-ended configuration. A single-ended stage, operating in non-linear areas, produces distortions. If the distorted pre-amplified signal is given to a push-pull stage, distortions are amplified as well.

A technique to reduce distortions, across all amplifier stages, is the *global negative feedback*. Global negative feedback consists in using a *negative feedback loop*, which subtracts the output signal, appropriately attenuated, from the input signal.

The effect of the global negative feedback can be explained intuitively as follows. If the input signal and the output signal have identical shapes, the subtraction has the only effect of attenuating the input signal to be amplified and consequently the produced output signal. Similarly, when distortion signal is fed back, and subtracted from the input signal, distortion is also reduced. Suppose now we compare two output signals having the same level, one produced with global negative feedback and the other without[3]. Note that distortion level depends on the output level, and does not depend on input level. This implies that, in both cases, the distortion internally introduced is the same. However, the outputs signal produced with negative feedback has less distortion, given that distortion itself has been attenuated through the feedback loop.

Figure 31 shows the basic schema of a global negative feedback loop. The feedback signal is produced by the voltage divider, composed of resistors R_1 and R_2, from the output signal. Resistor R_1 is generally called the *feedback resistor*. The feedback signal V_{fb}, provided by the voltage divider, is applied to the cathode of the input stage vacuum tube. When the output signal and the input signal have the same phase, this schema has the effect of subtracting the feedback signal V_{fb} from the input signal V_{in}. In fact, in this case, the feedback signal shifts the cathode voltage in the same direction of the input signal, reducing the grid to cathode voltage. In this way the signal at the grid of the input stage is $V_{in}^{fb} = V_{in} - V_{fb}$. The feedback signal V_{fb} is obtained using the voltage divider equation:

$$\boxed{V_{fb} = \frac{R_2}{R_1 + R_2} \cdot V_{out} = \beta \cdot V_{out}}$$

The factor β

$$\boxed{\beta = \frac{R_2}{R_1 + R_2}}$$

---

[3] In order to have the same output levels, when global negative feedback is used the input level should be appropriately increased.

is generally referred as the *feedback factor*. The amount of feedback can be set by choosing β, by way of the voltage divider resistors.

Note that, if the phase of the output signal were inverted, with respect the input signal, the circuit would produce a positive feedback, which increases distortions and produces oscillations. When phase shifts are produced through the amplifier's stages and affect the correct operation of the feedback loop, a capacitance C_{sn} can be used, together with resistor R_1, to form a *step network*, which has the purpose of maintaining a correct phase of the feedback signal at all relevant frequencies, guaranteeing stability of the amplifier. This is better discussed in Section 4.4.3.

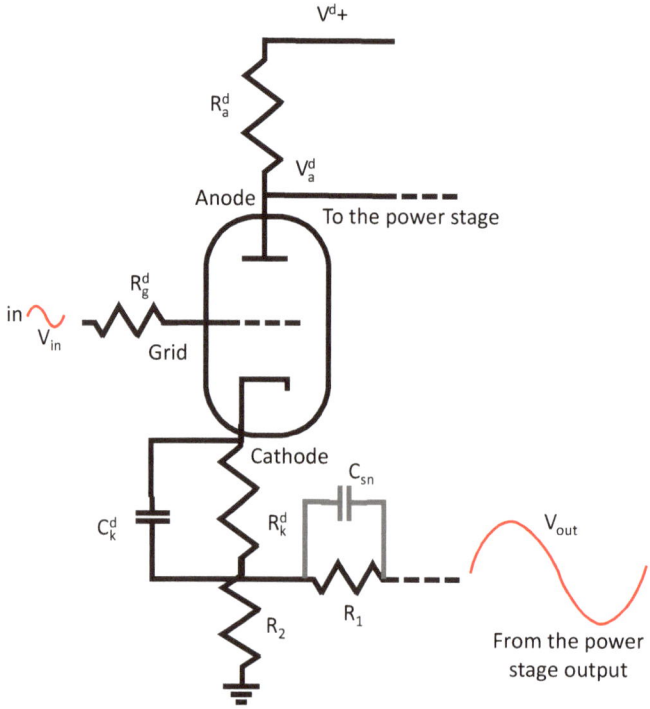

Figure 31: Basic schema for global negative feedback.
Global negative feedback is obtained by feeding the attenuated output signal back to the cathode of the input stage vacuum tube. If the output signal and the input signal have the same phase, the first is subtracted from the second, reducing distortions in the amplified signal. The voltage divider composed of the resistors R_1 and R_2 has the effect of attenuating the output voltage used as feedback. The capacitor C_{sn} forms, with resistor R_1, a step network to adjust the feedback signal phase to guarantee stability.

4.4.1 Gain with negative feedback

The gain of the amplifier without negative feedback is referred as the *open loop gain* of the amplifier. The open loop gain A is the ratio between the output and the input signals:

$$A = \frac{V_{out}}{V_{in}}.$$

We said that, when there is global negative feedback, the signal seen by the grid is $V_{in}^{fb} = V_{in} - V_{fb}$. Therefore, the output signal is

$$V_{out}^{fb} = A \cdot V_{in}^{fb} = A \cdot (V_{in} - V_{fb}) = A \cdot V_{in} - A \cdot \beta \cdot V_{out}^{fb}.$$

Simplifying we obtain

$$V_{out}^{fb} = A \cdot V_{in} \cdot \frac{1}{1 + A\beta}.$$

The *closed loop gain* A^{fb} of the amplifier, which is the gain of the amplifier when global negative feedback is used, can be obtained as the ratio between the output signal, with negative feedback loop, and the input signal:

$$\boxed{A^{fb} = A \frac{1}{1 + A\beta}.}$$

The quantity $A\beta$ is generally referred as the *loop gain*, which is the gain seen *in* the feedback loop.

It is generally useful expressing the amount of negative feedback applied as the reduction of the gain in the amplifier. This can be easily obtained by expressing the gain in dB and computing the amount of feedback fb_{dB} as:

$$\boxed{fb_{dB} = A_{db} - A_{db}^{fb} = 20 \cdot \log\left(\frac{A}{A^{fb}}\right) = 20 \cdot \log(1 + A\beta).}$$

For instance, if the gain of the amplifier without negative feedback is 20dB and the gain with global negative feedback is 14dB, we say that we apply fb_{dB}=6dB of feedback.

4.4.2 Benefits of negative feedback

The use of negative feedbacks has several advantages. It stabilizes the gain of the amplifier, decreases output impedance, increases input impedance, increases bandwidth, and reduces distortions.

In the following, as an example, we discuss how negative feedback reduces harmonic distortions[4] by a factor $1/(1+A\beta)$, at the same output level of the amplifier without negative feedback.

Suppose there is no negative feedback loop. In correspondence of the input signal V_{in}, the amplifier produces the output signal $V_{out} + V_d$, where V_d is the harmonic distortion introduced by the amplifier itself. The harmonic distortion percentage HD is measured as the ratio between the harmonic distortion signal and the output signal:

$$HD = \frac{V_d}{V_{out}}$$

Suppose now negative feedback loop is used and the input signal is increased from V_{in} to V'_{in}, to compensate the gain loss and obtain an output signal V_{out}^{fb} equal to the output signal obtained without negative feedback. In other words, the increased input signal V'_{in} is such that $V_{out}^{fb} = V_{out}$. Now, the output of the amplifier, considering also distortions, is $V_{out}^{fb} + V_d^{fb}$. We can treat separately these two components using the equations discussed before:

$$V_{out}^{fb} = AV'_{in} - A \cdot \beta \cdot V_{out}^{fb}$$

$$V_d^{fb} = V_d - A \cdot \beta \cdot V_d^{fb}.$$

---

[4] Harmonic Distortion and Negative Feedback in Audio-Frequency Amplifiers, Engineering Training Supplement, No. 3, Issue 2, British Broadcasting Corporation (BBC), December 1950

The harmonic distortion V_d depends only on the output signal V_{out} (without negative feedback) or V_{out}^{fb} (with negative feedback). Given that we set V_{out}^{fb} equal to the output signal without feedback V_{out}, the harmonic distortion V_d, internally introduced by the amplifier, is the same in both cases. However, the second of the above equations says that when using negative feedback, V_d is also fed back and attenuated to V_d^{fb}. Simplifying, as before, we obtain that

$$V_d^{fb} = V_d \frac{1}{1+A\beta}.$$

With negative feedback, the produced harmonic distortion V_d^{fb} is attenuated, with respect to the harmonic distortion V_d generated without negative feedback, at the same output level. For instance, a negative feedback f_{dB} of 20 dB implies a closed loop gain 10 times lower than the open loop gain. However, harmonic distortion, at the same output level, will be ten times lower as well.

4.4.3 Stability of negative feedback

The schema for negative feedback, given in Figure 31, requires that the input signal and the negative feedback signal, applied to the cathode, have the same phase. However, when the input signal goes through the amplifier stages, its phase might shift significantly, reaching in some cases a 180° phase-shift. With a 180° phase-shift, the feedback signal has an opposite phase with respect to the input signal and the feedback circuit becomes a positive feedback circuit. Positive feedback is dangerous because it might introduce instability and oscillations.

Consider that both low-pass and high-pass filters produce a 45° phase-shift at their cut-off frequency and that phase-shifts are accumulated in a sequence of filters. Therefore, a 180° phase-shift becomes probable at frequencies near the borders of the bandwidth of the amplifier.

Suppose f_{180} is the frequency where the 180° phase-shift occurs, and A_{180} is the amplifier gain at this frequency. If the loop gain $A_{180}\beta$ is smaller than 1 (that is $fb_{dB} < 6dB$), positive feedback introduces just a gain peak at

f_{180}. However, if $A_{180}\beta$ is greater or equal to 1 (that is $fb_{dB} \geq 6dB$), the amplifier will oscillate. In order to avoid this, the amplifier has to be designed so that the frequency, where the 180° phase-shifts occurs, is where $A_{180}\beta$ becomes smaller than 1. Low-pass filters, obtained using grid stopper resistors, and high pass filters, produced by inter-stage coupling capacitors, can accomplish to this task.

However, not always low-pass and high-pass filters are able to effectively eliminate the conditions for oscillation and instability. In these cases, a *step-network* can be used to move the 180° phase-shift where $A_{180}\beta$ is smaller than 1. A step network is obtained by using a capacitor in parallel with the feedback resistor, as shown in Figure 31, with capacitor C_{sn} and resistor R_1.

Example 15: Step network for negative feedback stability

Consider, for instance, an amplifier with an open loop gain A=500 (54dB). Suppose that the solid and dashed red plot in Figure 32 respectively give the gain and phase-shift of the amplifier, with respect to the frequency. We can see that the amplifier introduces a phase-shift of -180°, around 240K Hz, where the gain A_{240KHz} is still 25 (28dB). Suppose we apply a feedback factor β of 0.1, corresponding to a loop gain β·A=0.1·355=35.5 (fb_{dB}=31.2dB), by setting R_1 to 10K Ohms and R_2 to 1K Ohms. At 240K Hz we have that the loop gain is β·A_{240KHz}=0.1·25=2.5 (fb_{dB}=10.9dB). Since the loop gain at 240K Hz is greater than 1 (fb_{dB}>6dB) and the phase shift is -180°, the amplifier will oscillate.

To avoid oscillation, we have to move the -180° phase-shift at a frequency *f* where the loop gain βA_f is smaller than 1 (fb_{dB}<6dB). To have βA_f<1, with β=0.1, we need *f* such that the amplifier gain A_f at frequency *f* is A_f < 1/β=10 (20dB). The frequency where we have a 20dB gain is *f*=400K Hz. At this frequency, the phase-shift is -220°. To go below -180° we have to compensate at least 40°. Using a capacitor C_{sn} of 80 pF, in parallel with R_1, we obtain the frequency compensation shown by the red dashed line in Figure 33. We can see that the phase compensation at 240K Hz is 45° and at 400 Hz is 54°. The solid and dashed blue line in Figure 32 represents the loop gain and its phase-shift, with this compensation. At 240K Hz the phase-shift is now -135°, at 400K Hz it is -166°. The -180° phase-shift now occurs at 500K Hz, where the amplifier gain A_{500KHz} is 5 (14 dB), thus below the threshold 10, which we identified before. The amplifier now is stable, in fact, $\beta A_{500}=0.1 \cdot 5=0.5$, and $fb_{dB}=3.5dB$

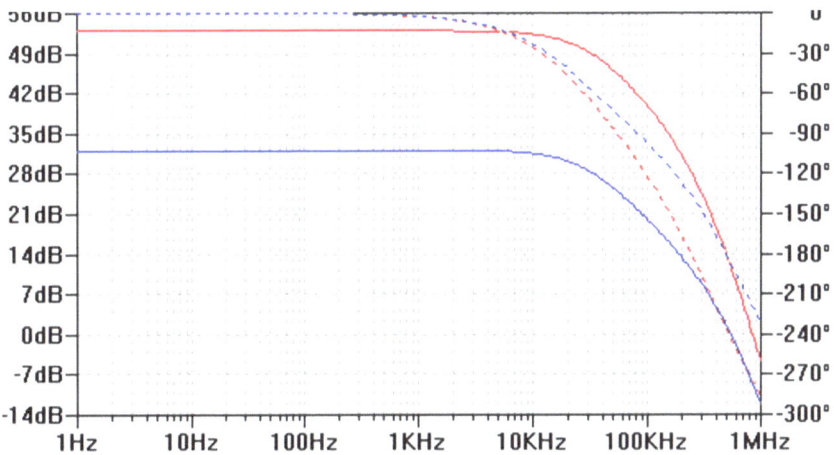

Figure 32: Phase compensation for stable negative feedback.
The gain and the phase-shift of the amplifier, with respect to the frequency, represented respectively by the red solid and dashed lines, show that we have a -180° phase-shift at 240K Hz, where the gain is still 28dB. Compensation, applied to the feedback loop, produces the loop gain and a phase-shift depicted by the blue solid and dashed lines. At 240K Hz, the phase-shift is 135°. 180° phase-shift is at 500K Hz, where amplifier gain is just 14dB.

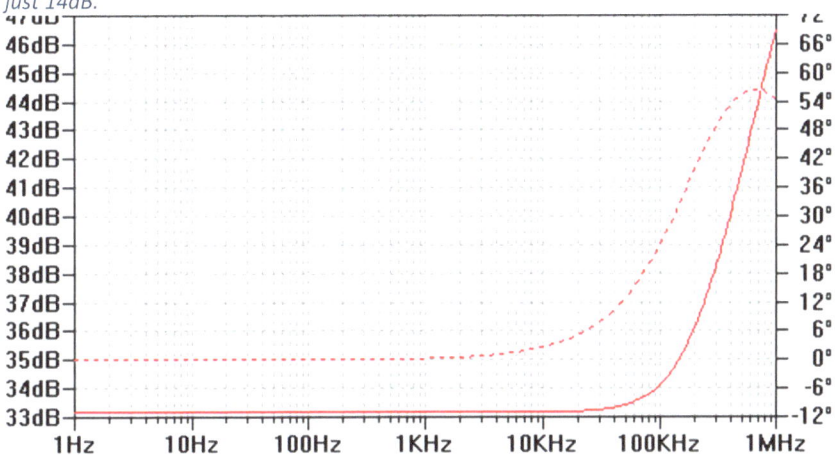

Figure 33: Phase compensation with a step network.
The step network, introduced in the feedback loop, produces the phase compensation depicted by the dashed line. We have a 45° compensation at 240K Hz and 54° at 400K Hz.

Chapter 5:
Power supply unit

The power supply unit has to provide all components of the amplifier with the needed voltage and current. The power supply unit is a very relevant component of the entire amplifier. A perfect design of all amplifier stages can be undermined by a poorly designed power supply. A good power supply, for a hi-fi amplifier, has to minimize hum and noise, generally introduced by the power supply rectifier, and has to provide a constant voltage, even in presence of transients requiring a sudden surplus of current.

The power supply unit has to supply electric power to the following parts:

- the (anodes of the) vacuum tubes in the various amplifier stages
- the vacuum tubes heaters
- the vacuum tubes grid, when fixed bias is used

In the following, we will discuss them separately.

5.1 Power supply for the amplifier stages

The different stages of the amplifier require different voltage, absorb a different amount of current, and are more or less sensitive to noise (e.g. hum) produced by the power supply itself. The power supply unit has to consider these differences.

The power supply unit is composed of a power *transformer*, followed by a *rectifier,* and a sequence of *smoothing filters* dedicated to the different stages, as depicted in Figure 34.

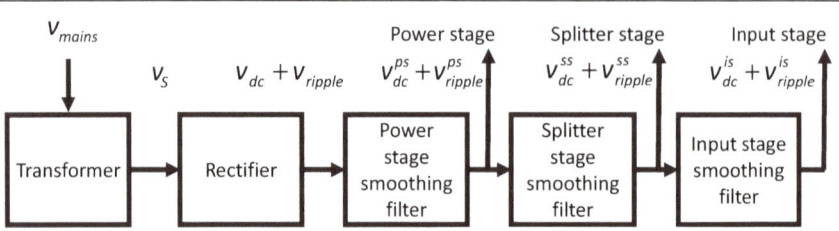

Figure 34: Basic components of a Power Supply Unit.
The power supply unit consists of chain of components containing a power supply transformer, a rectifier, and a sequence of smoothing filters.

The transformer takes as input, at its primary winding, the mains AC V_{mains} and returns, from its secondary winding, the AC voltage V_s to be given to the rectifier. The rectifier converts the received AC voltage into a DC voltage V_{dc} plus a residual AC voltage V_{ripple} called the *ripple voltage*. This is due to the fact that there is no ideal rectifier, and a residual AC ripple voltage always remains on top of the wanted DC voltage. The sequence of smoothing filters, following the rectifier, have the purpose of both reducing the DC voltage, to the value needed by the corresponding stage, and reducing the ripple voltage, to a value tolerated by the stage itself.

In the following, we will first discuss rectifiers circuits, then we will discuss smoothing filters.

5.1.1 Rectifiers

Mains voltage has to be adapted to what needed by the vacuum tubes. For instance, mains voltage in Europe is 230V. This is generally not enough for most tubes, which often need more voltage. In addition, mains voltage is AC, while vacuum tubes require DC. Therefore, a step-up power transformer is first needed to bring the mains voltage up to the needed voltage. Then, the rectifier converts the AC current produced by the transformer into a DC current.

Figure 35 shows the schema of three very common types of transformers and rectifiers combinations. In the figure, R_L represents the load of the power supply.

- The *half wave* rectifier schema, at the top of the figure, rectifies the AC voltage V_S produced by the secondary of the transformer, using a single diode. The diode conducts only during the positive half cycles of V_S. Therefore, the voltage waveform, produced by this rectifier, has the same shape than V_S during the positive half cycles and it is zero during the negative half cycles.
- The *full wave rectifier* uses a centre-tapped transformer and two diodes. The centre tap is connected to ground. The two diodes have common cathodes, and anodes connected to the two ends of the transformer. The AC voltage V_S, between the centre-tap and each end of the transformer, is half than the voltage measured between the two ends themselves. The phases of the voltage measured between the centre tap and one transformer end is the inverse of the phase from the centre tap and the other end. Therefore, when one diode sees a positive half cycle, the other sees a negative half cycle, and vice versa. The result is that the output waveform has positive pulses during all half cycles.
- The *full wave bridge rectifier* uses four diodes to convert all half cycles of the AC voltage V_S into positive pulses. The output waveform is the same that that produced by the full wave rectifier.

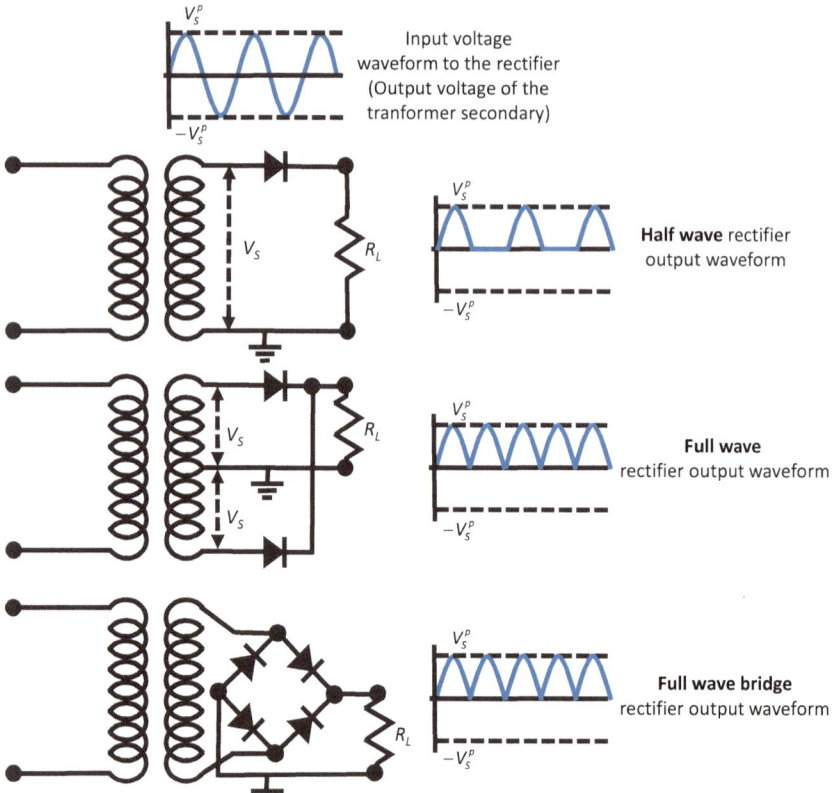

Figure 35: Half wave, full wave, and full wave bridge rectifiers.
The power transformer receives at its primary the mains AC voltage and produces, from its secondary, an AC output voltage V_S. The rectifier eliminates negative pulses. R_L represents the load of the power supply.
The half wave rectifier, on top, just conducts during the positive half cycles. Therefore, the output voltage waveform has the same shape of V_S, during the positive half cycles. It is 0 during the negative half cycles. The full wave rectifier, in the middle, uses a centre-tapped transformer. The AC output voltage V_S is measured from the centre tap to the two ends. The output waveform has positive pulses during all half cycles. The full wave bridge rectifier does not need a centre-tapped transformer. It uses a bridge to convert the negative half cycles into positive. The resulting waveform is the same than that of the full wave rectifier.

All types of rectifier eliminate negative voltage coming from the V_S waveform. However, the output voltage has a pulse waveform with a significant AC ripple component. The AC ripple voltage has a frequency equal to the mains frequency, for the half wave rectifier, and twice the mains frequency for the full wave rectifiers. The rectified voltage varies between zero and the peak voltage V_S^p. The peak voltage V_S^p, reached by the pulses, is equal to the peak of the AC voltage V_s. If V_s is given as RMS voltage, then the peek voltage is $V_s^p = 1.414 \cdot V_s$.

The AC ripple voltage introduces in the output signal, produced by the amplifier, an unacceptable humming noise. A steadier DC voltage is needed and can be obtained by placing, after the rectifier, a reservoir capacitor and by using a sequence of smoothing filters, as discussed in next sections.

5.1.2 Reservoir capacitor

A very important component to complete a rectifier is the *reservoir capacitor C_R* connected between the positive and the ground, as depicted in Figure 36. It significantly reduces the ripple voltage and returns a steadier DC voltage.

We explain the usage of the reservoir capacitor using a full wave rectifier. However, this discussion can also be generalized to other types of rectifiers.

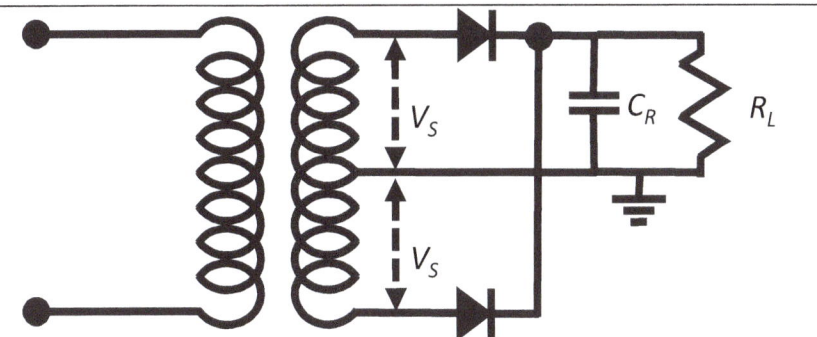

Figure 36: The reservoir capacitor.
The reservoir capacitor C_R connected between the positive and the ground significantly reduces the ripple of the rectified voltage.

The effect of the reservoir capacitor is shown in Figure 37. The blue waveform represents the rectifier output voltage, when no reservoir capacitor is used, and no load is connected to the power supply. The red dotted waveform represents the positive pulse voltage of the two halves of the transformer secondary, with the reservoir capacitor and a load. The capacitor initially charges almost up to the peak voltage V_s^p, as depicted by the red solid waveform. When the pulse voltage of the transformer secondary (red dotted waveform) decrease below the capacitor voltage, current no longer traverses the diode. At this point, the capacitor feeds the load and slowly discharges. When the next pulse voltage is higher than the capacitor voltage, an intense current peak traverses the diode and the capacitor charges again quickly. The resulting voltage has a saw-tooth waveform. It goes quickly up during the charging phases. It goes slowly down during the discharging phases. The frequency is the same as the frequency of pulses arriving from the rectifier.

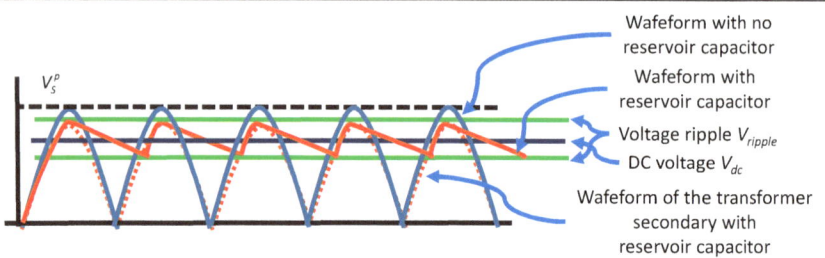

Figure 37: Ripple voltage reduction with reservoir capacitor.
The reservoir capacitor is charged almost up to the peak voltage during each half cycle. Its voltage is depicted by the solid red waveform. The voltage produced by the transformer secondary with a reservoir capacitor is represented by the red dotted waveform. When this voltage is below the reservoir capacitor voltage, the capacitor feeds the load and start discharging. When the rectified voltage is again high enough, it recharges the reservoir capacitor. The voltage of the reservoir capacitor has saw-tooth like waveform. It goes quickly up during the charging phases. It goes slowly down during the discharging phases.

The figure shows that the peak voltage reached by the capacitor, as depicted by the red solid waveform, is lower than the peak voltage V_s^p of the rectifier without reservoir capacitor and load. This depends on the speed at which the capacitor charges, which in turns depends on the capacitor capacitance, the load resistance, and the transformer impedance.

With the reservoir capacitor a ripple voltage still occurs, even if it is much smaller than that produced by the rectifier alone. The ripple voltage is due to the charging and discharging phases of the reservoir capacitor. On one hand, charging depends on the transformer output impedance and reservoir capacitance. Low output transformer impedance and low capacitance increase the peak voltage ripple and reduce reservoir capacitor charging time. On the other hand, reservoir capacitor discharging depends on the load impedance, the ripple frequency, and again the reservoir capacitance. Large load impedance, high ripple frequency, and large reservoir capacitance reduce the discharging voltage drop.

The output voltage is the sum of a DC voltage plus an AC (saw-tooth) ripple voltage $V_{dc}+V_{ripple}$. Both can be estimated with sufficient accuracy, using results of a study carried out by Shade[5], discussed later. However, in order to estimate the DC output voltage and ripple voltage, we first need to estimate the *transformer output impedance* and the *load impedance*. Obviously, the load impedance represents the impedance offered by the amplifier.

5.1.3 Transformer output impedance

The transformer output impedance R_s can be made explicit using an equivalent circuit where we place two resistors R_S at the two transformer's ends, as in Figure 38. Two components contribute to the resistance R_s. The first, $R_{sec\text{-}wind}$, is the secondary winding resistance. The second, $R_{prim\text{-}wind}$, is the primary winding resistance, reflected to the secondary. Since we are using a centre-tapped transformer, where contribution to form the output voltage is given by one transformer section at time, we need to consider as $R_{sec\text{-}wind}$ just the resistance between the tap and one transformer end. The primary winding resistance, reflected to the secondary is equal to the primary winding resistance times the square of the ratio between the output voltage V_S and input voltage V_{mains} (see Section 4.1.2 for discussion on the impedance reflected by transformers). R_S is obtained as the sum of these two components:

$$\boxed{R_S = R_{sec-wind} + R_{prim-wind} \cdot \left(\frac{V_S}{V_{mains}}\right)^2}.$$

V_S is the voltage between the centre tap and one transformer end.

---

[5] O.H. Shade, "Analysis of Rectifier Operations", Proceedings of I.R.E, August 1943, pp. 341-361.

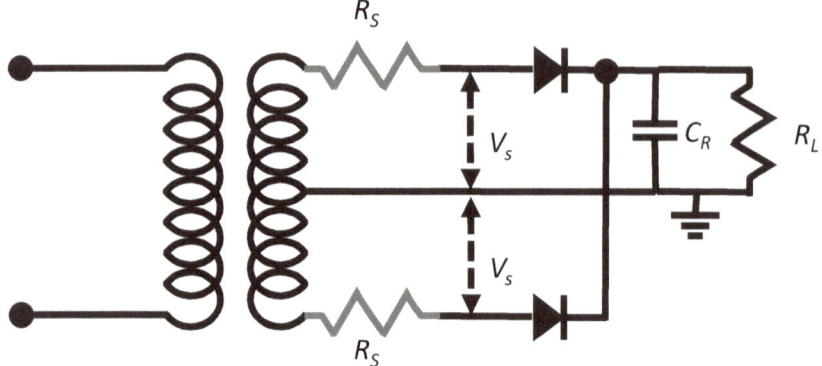

Figure 38: Transformer output resistance made explicit in an equivalent circuit.
To make the transformer resistance explicit we draw an equivalent circuit where two resistors R_S connect the two transformer's ends to the two diodes. R_S includes the secondary winding resistance and the primary winding resistance reflected to the primary.

Example 16: Power supply transformer output impedance

Suppose for instance the primary winding resistance is 4 Ohms, the secondary winding resistance is 20 Ohms, the mains voltage is 230V, and the secondary transformer output voltage V_s is 325V. We have that

$$R_S = 20 ohm + 4 ohm \cdot \left(\frac{325V}{230V}\right)^2 \approx 28 ohm.$$

5.1.4 Load estimation

The load R_L is the resistance seen by the reservoir capacitor, which is the impedance offered to the power supply by all amplifier stages working in parallel. The impedance in each stage is the sum of the smoothing filter impedance, of that stage, plus the stage impedance itself. Provided the power stage is the first stage, the load R_L can be roughly estimated, using the Ohm's law, as the ratio between the voltage required by the power stage and the sum of the current absorbed by all stages. This approximation does not take into consideration the impedance of the power stage smoothing filter, which as discussed in Section 5.1.8, can only be computed once we know the DC output voltage of the rectifier. However, if the DC output voltage produced by the rectifier is not significantly higher than the voltage required by the power stage, the corresponding smoothing filter impedance is small, and the load seen by the reservoir does not significantly differ from this estimation.

Example 17: Impedance offered by the amplifier to the power supply

Suppose the power stage requires 400V and absorbs 80 mA, the phase splitter absorbs 1 mA, the input stage, also 1 mA. We have that

$$R_L \approx \left(\frac{400V}{80mA + 1mA + 1mA}\right) \approx 4.9kOhm.$$

If we have a stereo amplifier, we have to divide this by 2, since the amplifier has double current absorption.

5.1.5 Estimation of the DC output voltage

The plot in Figure 39, taken by Shade's work[6], puts in relationships all relevant variables and allows estimating the DC voltage V_{dc} for a full wave rectifier. Shade's work reports the same graphs also for the other types of rectifiers. Every red plot corresponds to a percentage of the transformer impedance R_S, with respect to the load resistance R_L. Knowing the mains frequency f_M, the reservoir capacitance C_R, and the load R_L, we fix $2\pi f_M C_R R_L$ on the horizontal axis, and then we read the DC voltage V_{dc} as a percentage of the peak transformer voltage V_S^p, on the corresponding red plot.

Example 18: DC voltage output of a full wave rectifier

Suppose for instance the load R_L is 2.5K Ohms, the reservoir capacitor C_R is 22 µF, R_S is 28 Ohms, the mains frequency f_M is 50 Hz, and the secondary voltage is V_S=325V. The ratio R_S/R_L is 1.1% and $2\pi f_M C_R R_L$=17.2. Using the plot corresponding to a percentage of 1, in correspondence of 17.2 we have a percentage V_{dc}/V_S^p around 92%. Since the peak voltage is $V_S^p = 1,414 \cdot V_S \approx 460V$, we have that the estimated DC output voltage is $V_{dc} = 460V \cdot 92\% \approx 423V$.

---

[6] Note that Shade's works considers vacuum tube diodes and includes their resistance in the estimation of R_L. Here we are considering solid-state diodes, and given the high voltages at hand, their resistance can be ignored.

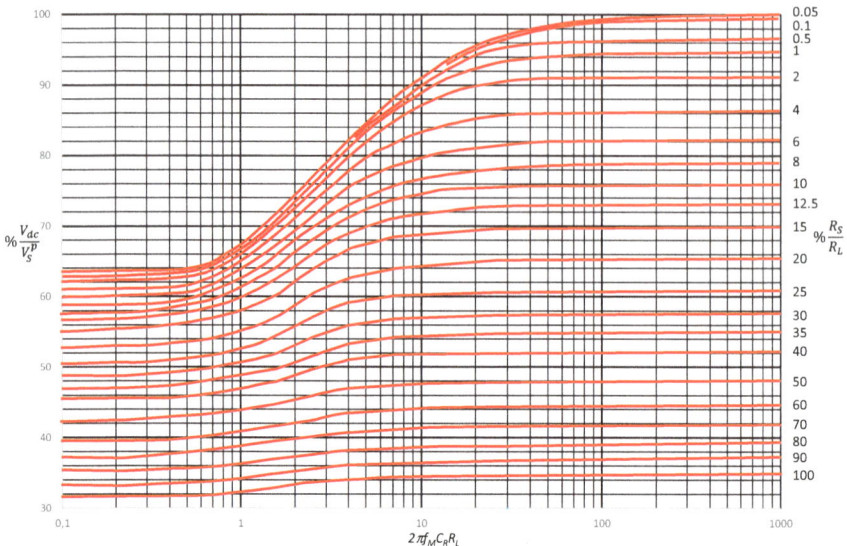

*Figure 39: Determining the output DC voltage of a full wave rectifier.
Relationships between output DC voltage, secondary transformer peak voltage, reservoir capacitance, mains frequency, load, and transformer impedance for a full wave rectifier. Every red curve corresponds to the percentage between R_S and R_L. The load R_L is the impedance offered by the amplifier and the smoothing filters to the power supply. For each curve, knowing the mains frequency, the reservoir capacitance, and the load, we can obtain the ratio between the output DC voltage and the secondary transformer peak voltage.*

5.1.6 Estimation of the ripple voltage

The plot in Figure 40, also taken by Shade's work, allows us to estimate the output ripple voltage V_{ripple}. As before, every plot corresponds to a different percentage ratio between the transformer impedance R_S and the load resistance R_L. Using the mains frequency f_M, the reservoir capacitance C_R, and the load R_L, using one of the plots, we can obtain the ratio between the ripple voltage V_{ripple} and the DC output voltage V_{dc}

> **Example 19: Ripple voltage of a full wave rectifier**
>
> Assuming the parameters obtained in Example 18, the plot corresponding to a ratio of 1, in correspondence of 17.2, gives us a percentage V_{ripple}/V_{dc} around 4%. Therefore, the estimated ripple voltage is $V_{ripple} = 423V \cdot 4\% \approx 17V$.

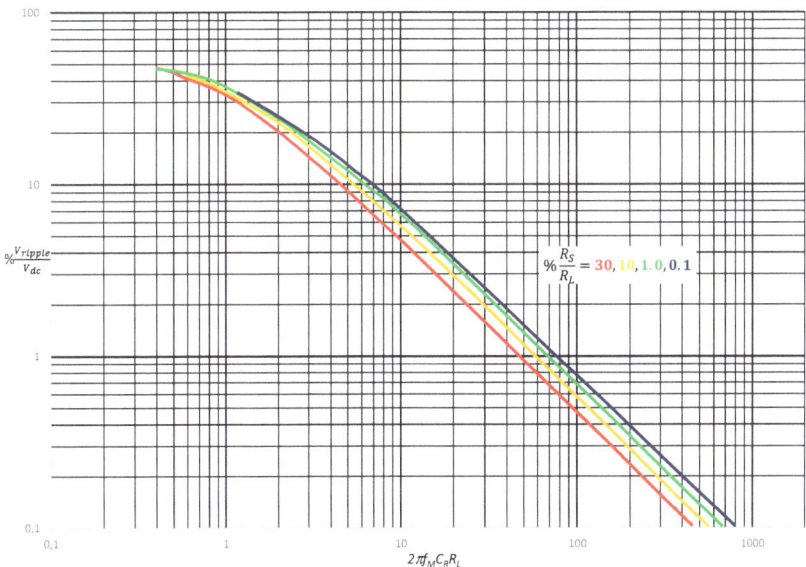

Figure 40: Determining the ripple voltage of a full wave rectifier. Relationships between output DC voltage, ripple voltage, reservoir capacitance, mains frequency, and load.

5.1.7 Estimation of the transformer secondary RMS current

Let us consider again, for this discussion, a full wave rectifier. Let I_L be the current absorbed by the load. The current is fed, in turns, by the two sections of the transformer secondary through the two diodes. The average current I_{avg} flowing through each of the two transformer secondary sections and the two diodes is half the current traversing the load: $I_{avg} = I_L/2$. However, previously we said that the transformer delivers current in intense peaks, during the charging phases of the reservoir capacitor. In fact, when the voltage of one of the transformer secondary sections is higher than the reservoir capacitor, the capacitor is quickly recharged with intense burst of current delivered by the transformer secondary through the two diodes, with a frequency twice the mains frequency. When the voltage of the transformer secondary section goes below the reservoir capacitor voltage, the diode does not conduce and the capacitor discharges, until the voltage of the other transformer secondary section is again higher than that of the reservoir capacitor, and so on. It is not easy to guess the RMS current I_S traversing the two sections of the transformer secondary and the two diodes, given these current bursts.

Shade's work also reports some guidelines for estimating the RMS current I_S traversing the secondary winding of the power supply transformer and the rectifier diodes. The graph in Figure 41, derived by an equivalent graph from Shade's work, puts in relationships all variable at hand, already used before. As before, every red plot corresponds to the ratio between the transformer impedance R_S and the load resistance R_L. Knowing the mains frequency f_M, the reservoir capacitance C_R, and the load R_L, using one of the red plots, we can obtain the ratio between the RMS current I_S and the current I_L absorbed by the load.

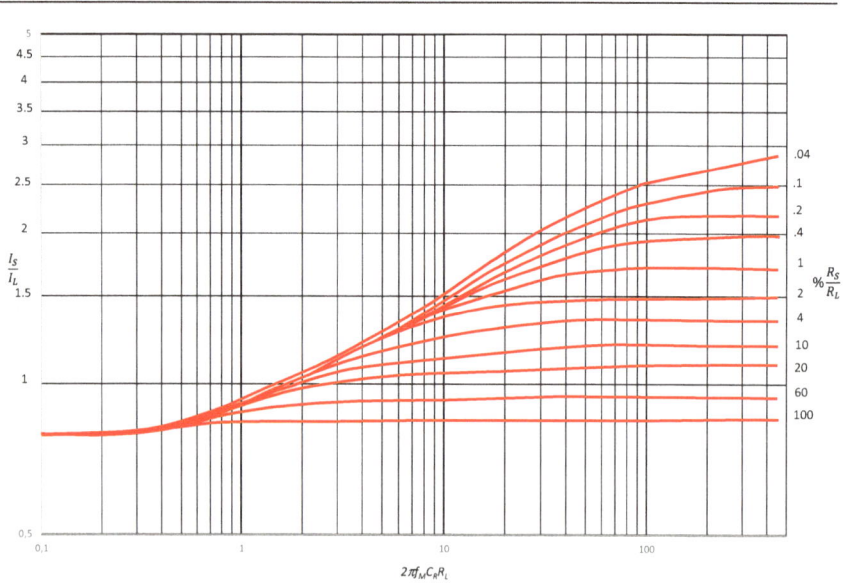

Figure 41: Transformer secondary RMS current of a full wave rectifier.
Relationships between output DC current, transformer secondary RMS current, reservoir capacitance, mains frequency, and load.

Example 20: RMS current in a transformer secondary of a full wave rectifier

Let us use again the parameters obtained in Example 18. The plot corresponding to a ratio of 1, in correspondence of 17.2, gives us a percentage I_S/I_L around 1.5. In Example 18, the estimated output voltage was 423V. With a load of 2.5K Ohms the current is I_L=423V/2.5K Ohms=170mA. The RMS current traversing the two transformer secondary sections and diodes is I_S =1.5·I_L=1.5·170mA=255mA.

The estimated value for I_S can be used for determining the current rating of both the transformer and the diodes. Transformer and diodes should be chosen with a current rating higher than the estimated value, to guarantee safe operation also on extreme conditions. Generally, transformer and diode rating are chosen something like twice the obtained value. For instance, according to the value determined in Example 18 and Example 20, we can chose a centre tapped transformer rated for 650 V (325 V each section) at 500 mA (or simply 650V C.T. @ 500mA).

5.1.8 Smoothing filters

Smoothing filters are used to reduce DC voltage to that needed by each individual stage, and to further reduce the ripple voltage to a value tolerated by each stage. The initial stages tolerate much less ripple than the final stages. Reasonable values are[7]:

- Input stage: 0.001%-0.002%
- Phase splitter stage: 0.01%-0.05%
- Push-Pull power stage: 0.5%-2%

A smoothing filter is in practice a low pass filter. It can be obtained using an inductor-capacitor network (LC low pass filter) or a resistor-capacitor network (RC low pass filter). Here we will discuss how a smoothing filter can be obtained using an RC network. What we discuss here can be easily extended to the case of an LC smoothing filter.

A simple schema for an RC smoothing filter is given in Figure 42. If this is the first smoothing filter, its input voltage arrives from the reservoir capacitor. Elsewhere, its input comes from the preceding smoothing

---

[7] "Power Supply for High-End Vacuum Tube Amplifiers", Eugene Karpov (in Russian: "Источники питания для ламповой High-End аудио аппаратуры", Евгений Карпов)
"http://next-tube.com/articles.php?sub_menu_item=99&article=articles/supply/SupplyRu.inc"   -
(Link checked on August 2018)

filter. In both cases, the input voltage consists of a DC voltage V_{dc}^i plus an AC ripple voltage V_{ripple}^i. The DC voltage is only affected by the resistor R_{flt} of the filter. The combined action of the resistor R_{flt} and the capacitor C_{flt} has effect on the AC ripple voltage.

Let us suppose that the stage of the amplifier powered by the filter (Stage 2 in Figure 42) requires a DC voltage V_2 and absorbs a current I_2. Let us also suppose that, the input DC voltage of the filter, or alternatively the DC voltage required by the previous amplifier stage (Stage 1 in Figure 42), is V_1. Of course, V_1 must be higher than V_2. Finally, let us suppose that the next stages, powered by the power supply, absorb a current I_{next}.

The resistor R_{flt} has to produce a DC voltage drop of $V_1 - V_2$. The current that traverses the resistor is the current absorbed by the Stage 2 plus that absorbed by the next stages, which is $I_2 + I_{next}$. To calculate the resistance of R_{flt} we can use the Ohm's law:

$$\boxed{R_{flt} = \frac{V_1 - V_2}{I_2 + I_{next}}}.$$

Example 21: Determining resistor for a power supply RC Smoothing filter

Suppose, for instance, Stage 1 is the phase splitter stage, Stage 2 is the input stage, and no other stages are powered after the input stage. Suppose the voltage needed by the phase splitter is V_1=380V, and the voltage needed by the input stage is V_2=300V. Suppose the current in the input stage is 0.9mA, which is also the total current traversing the resistor, given that no other stages are powered next. In this case, we obtain that the resistor value must be R_{flt}=(380V-300V)/0.9mA=89K Ohms. The closest standard is 82K Ohms and the dissipated power is P=(380V-300V)·0.9mA=0,072W.

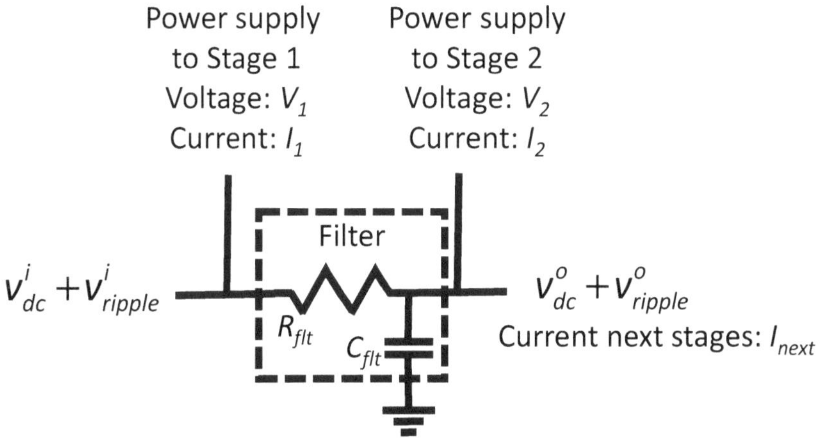

Figure 42: Power supply smoothing filters.
A smoothing filter takes as input a DC voltage plus a ripple voltage and produces a reduced DC voltage with reduced ripple voltage as well. The input can be produced by the rectifier or by a proceeding smoothing filter. The output is used to give electric power to an amplifier stage. The input is generally also used to supply power to another stage of the amplifier.

Let us now consider the ripple voltage. The resistor R_{flt} and the capacitor C_{flt} form a voltage divider for AC voltage. Given that the capacitor is a reactive component we have that its impedance is

$$X_{flt} = \frac{1}{2\pi \cdot f \cdot C_{flt}},$$

where f is the frequency of the ripple voltage. Remember that, for a full wave rectifier, the ripple frequency f is twice the mains frequency.

Using the reactive voltage divider equation, we obtain

$$\boxed{V_{ripple}^o = V_{ripple}^i \frac{X_{flt}}{\sqrt{X_{flt}^2 + R_{flt}^2}} = V_{ripple}^i \frac{1}{\sqrt{1 + (2\pi \cdot R_{flt} \cdot C_{flt} \cdot 2f_M)^2}}.}$$

Example 22: Determining voltage ripple of a power supply RC Smoothing filter

Suppose V^i_{ripple}=90µV, R_{flt}=82K Ohms and C_{flt}=22µF. In Europe, the mains frequency is 50 Hz, so the ripple frequency of a full wave rectifier is 100Hz. Therefore, we have that

$$V^o_{ripple} = 90\mu V \frac{1}{\sqrt{1+(2\pi \cdot 22Kohm \cdot 22\mu F \cdot 100Hz)^2}} \approx 0.08\mu V.$$

5.2 Power supply for the vacuum tube filaments

The thermionic effect takes place when the cathode is heated at a very high temperature so that electrons start leaving the cathode surface. If the anode voltage is higher than the cathode, electrons emitted by the cathode are attracted by the anode, and current starts flowing from the anode to the cathode.

There are two methods to heat the cathode: the cathode can be *directly heated* or *indirectly heated*. In vacuum tubes with directly heated cathode, the heating filament also plays the role of the cathode, and electrons are emitted by the filament itself. In this case, DC voltage has to be used to heat the filament, in order not to introduce hum in the amplified signals. In vacuum tubes with indirectly heated cathodes, the filament and the cathode are two distinct components. The cathode is a metal cylinder, or some other shape, which surrounds the filament. The filament, in this case, is electrically isolated from the cathode. This allows AC voltage to be used to heat the filament.

Many vacuum tubes, used in audio applications, have indirectly heated cathodes. Here we will discuss only this heating method and we will consider the use AC voltage to supply power to the filaments.

Many vacuum tubes require a filament voltage of 6.3V, which can be obtained by connecting the filament to a step-down transformer, as shown in Figure 43 a). Some vacuum tubes, as for instance 12AX7 vacuum tubes, are internally composed of two different vacuum tubes. In this case, each internal filament be can be either powered with 6.3V or can be connected in series and powered with a voltage of 12.6V, as shown in Figure 43 b).

When choosing the power supply transformer, in addition to the output voltage, the current rating should be considered as well. Vacuum tubes datasheets generally report the current absorbed by the heaters.

> **Example 23: Choosing heaters transformers**
>
> An EL34 vacuum tube requires 6.3V at 1.5A. If we need to provide electric power to the heaters of a stereo push-pull amplifier, composed of four EL34, with filaments connected in parallel, we need a step-down transformer of 6.3V, rated at least for 6A. 12AX7 tubes contains two vacuum tubes each requiring 6.3V at 0.15A. Suppose we use two of them (thus, four vacuum tubes) for both the input and phase splitter stages. In this case, we have two options. We can power all of them in parallel, as in Figure 43 a), with a transformer of 6.3V rated for at least 0.6A. Alternatively, we connect the two internal heaters of each 12AX7 in series and then connect the two series in parallel, as in Figure 43 b), with a transformer of 12.6V rated for at least 0.3A.

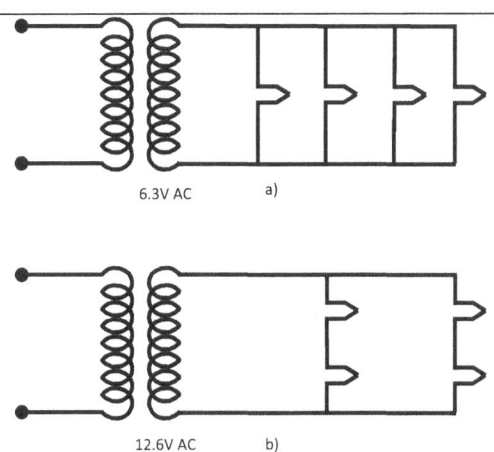

Figure 43: Heater's power supply.
Many vacuum tubes heaters requires 6.3V. All vacuum tubes heaters can be simply connected in parallel to a 6.3V transformer as show in a). Some vacuum tubes internally contain two distinct vacuum tubes, whose filament can be individually powered at 6.3V, or can be powered in series at 12.6V, as show in b).

5.2.1 Improved circuit: common and elevated voltage reference

Using the schemas given in Figure 43, the heater filaments are isolated from the other amplifier components. This implies that there is no common voltage reference and the filament voltage is floating, with respect to the voltage of the other amplifier components. The voltage difference between filament and cathode can assume uncontrolled large values. This can introduce hum in the audio signal, given to current leakage from the filament to the cathode. More importantly, an arc can occur between the filament and the cathode, which would damage the vacuum tubes.

Vacuum tubes datasheets generally report the maximum allowed positive and negative voltage between filament and cathode. This limit must be observed, to guarantee a safe operation of the vacuum tube. For instance, the Philips EL34 datasheet specify 100V as maximum filament to cathode voltage.

Generally, the voltage reference of the heater circuit is set higher than the cathode voltage, to reduce induced hum by eliminating current leakage between the filament and the cathode. In addition, in some configurations, cathode voltage is much higher that ground. Consider for instance a concertina phase splitter, where cathode voltage can be around 100V. In these cases, filament voltage must be elevated to be closer to the cathode voltage, and within the limits reported in the datasheets of the vacuum tubes.

Voltage reference can be set with a circuit like the one in Figure 44 a). A voltage divider composed of resistors R_1 and R_2, from high-tension V+ and ground, produces the needed elevated voltage reference. The voltage reference can be given to the filament power supply circuit using the centre tap of a centre tapped transformer. Note that, there is no current flowing from the filament power supply circuit to the voltage divider. The connection, from the voltage divider and the centre tap, just sets a common voltage reference. A smoothing capacitor C_{s2} is generally used in the voltage divider, to attenuate voltage ripple and other interferences. Large capacitance values, as for instance 1µF, or even 10µF, are generally used for the smoothing capacitor C_{s2}.

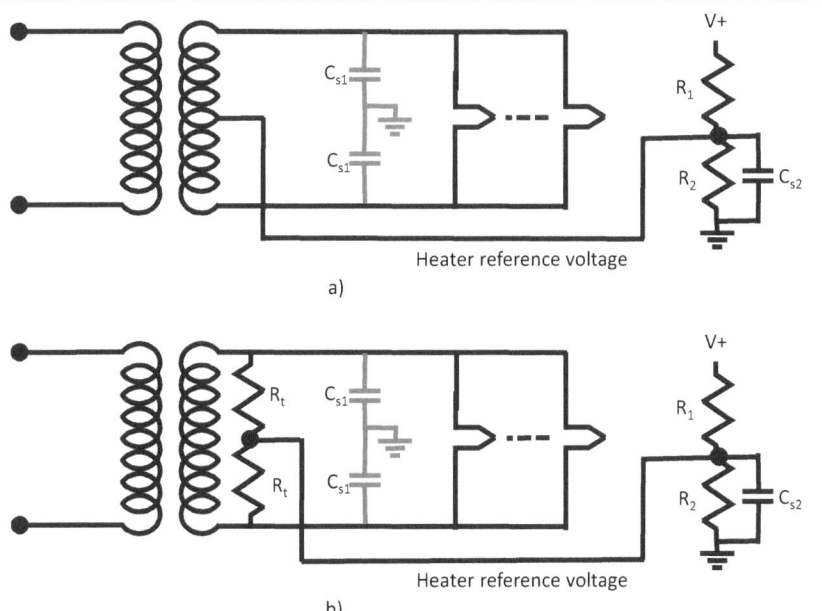

Figure 44: Heater's power supply elevated voltage reference
Voltage reference between filament and cathode can be set using a voltage divider, from V+ to ground, and feeding it through a centre tapped transformer as shown in a). In case the transformer does not have a centre tap, an artificial centre tap can be obtained connecting two appropriate resistors R_t between the two transformer ends, and feeding the voltage reference through them, as shown in b). It is generally convenient to decouple the lower resistor of the voltage divider with a large capacitor C_{s2}. Sometimes, it is also convenient to use the two capacitors C_{s1} to short to ground current peaks due, for instance, to solid-state rectifier switching interferences, which can transfer electromagnetic noise to the cathode.

Generally, R_1 and R_2 have large values, in order not to have too much current in the voltage divider. However, vacuum tube datasheets generally specify also the maximum filament to cathode resistance that can be tolerated, in addition to the above-mentioned filament to cathode voltage. This poses a limit to the maximum allowed value of R_2. For instance, the Philips EL34 datasheet specify 20K Ohms as maximum filament to cathode resistance.

Example 24: Heater voltage elevation

Suppose, we have an EL34 configured in fixed bias, so that its cathode is connected to ground. Suppose that the anode voltage V+ is 400V. We already said that Philips EL34 datasheets specify that maximum filament to cathode voltage is 100V and that maximum filament to cathode resistance is 20K Ohms. Accordingly, suppose that we want to set the filament to cathode voltage V_{fk} to 30V and that we want to have a filament to cathode resistance, corresponding to R_2, equal to 15K Ohms. Using the voltage divider equation, we have that

$$R_1 = \frac{R_2 \cdot (V_+ - V_{fk})}{V_{fk}} = \frac{15K\,Ohms \cdot (400V - 30V)}{30V} = 185K\,Ohms.$$

The closest standard is R_1=180K Ohms. The current dissipated by the voltage divider is 400V/(15K Ohms+180K Ohms)=2.05mA. The power dissipated by the two resistors R_1 and R_2 is respectively $(2.05mA)^2 \cdot 180K$ Ohms=0.76 W and $(2.05mA)^2 \cdot 15K$ Ohms =0.06 W.

5.2.2 Artificial transformer centre tap

Sometimes, heater transformers do not have a centre tap. In these cases, an artificial centre tap can be easily created by connecting two resistors R_t to the two transformer ends, as shown in Figure 44 b). The voltage reference, produced by the voltage divider, can be fed through these two resistors, as shown in the figure. Note that this resistance should be added to the heater voltage elevation circuit resistance and should be taken into consideration when checking that the maximum cathode to filament resistance is not exceeded.

Generally, the two resistors R_t have a resistance around 100 or 220 Ohms. As we said previously, there is no current flowing from the voltage divider to the filament power supply circuit. However, current flows across the two resistors R_t, from one transformer end to the other. Suppose we use two 100 Ohms resistors. The current is 6.3V/200 Ohms=31.5mA and the dissipated power is $(31.5mA)^2 \cdot 200$ Ohms=0.2W.

Sometimes, a potentiometer is used in place of the two resistors R_t and the reference voltage produced by the voltage divider is fed through the potentiometer wiper terminal. The potentiometer allows varying the values of the resistance between the artificial centre tap and the two transformer ends, to find the position where the hum, possibly introduced by the filament power supply circuit, is minimized.

5.2.3 Reducing electromagnetic interferences from other transformer windings

In many cases, instead of using several power supply transformers, one single transformer is used, which contains several independent windings for the various amplifier components. One of these is the heater power supply winding. In these cases, there could be interferences between one section and the others. For instance, consider that very short and intense current impulses go through the transformer windings connected to solid state rectifiers, when using large values of reservoir capacitors. Impulses have 50Hz or 100Hz frequency, depending on the type of rectifier. They are very short and intense, and also contain very high frequency components that produce electromagnetic interferences that may travel across the various sections of the transformer. When these interferences reach the filament winding, they can also interfere with the cathode and can be heard as a buzzing noise. The interferences can be significantly reduced, by shorting them to ground, using low value capacitors connected from the two ends, of the heaters' transformer, to ground. This is depicted by the two greyed C_{s1} capacitors in Figure 44 a). Small capacitance values, for instance 6.8 nF, can be used to accomplish to this task.

5.3 Power supply for the fixed grid bias

The grid of a vacuum tube needs to be negative with respect to the cathode. When fixed bias is used, as explained in Section 3.6.1, the cathode is connected to ground and a separate power supply is needed to produce the needed negative grid bias voltage.

Figure 45 reports a basic schema to produce the negative grid voltage bias. Required voltage is generally lower than 100V. Therefore, a step-down transformer is used to reduce the mains voltage to a voltage closer to the needed one. In this discussion, we use a bridge rectifier. However, the other rectifier options, discussed in Section 5.1.1, can be used as well. In particular, note that a very limited current flows in a well-designed grid bias circuit. Consequently, in some cases, also a half wave rectifier is sufficient to have a very limited ripple, further simplifying the overall schema.

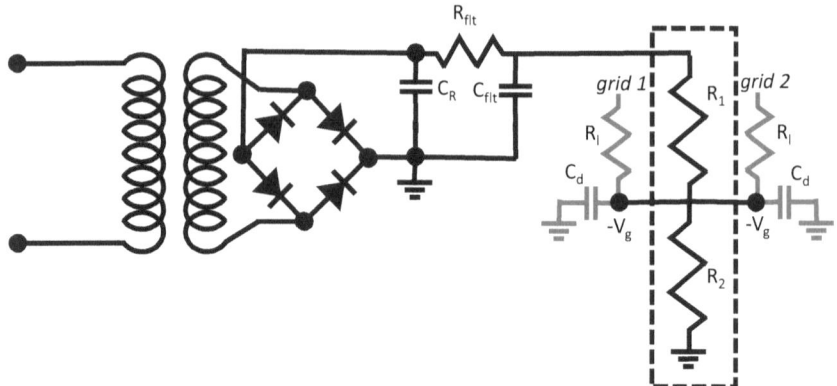

Figure 45: Power supply for grid bias
Negative voltage with respect to ground is obtained by connecting the positive terminal of the rectifier to ground. In this way, the other terminal is negative with respect to ground. As usual, a reservoir capacitor and a smoothing filter follow the rectifier, to reduce the ripple voltage. The wanted grid bias voltage $-V_g$ is obtained with the voltage divider composed of resistors R_1 and R_2. Grid bias voltage is fed to the grids of the two push-pull vacuum tubes through their grid leak resistors R_l. In order to avoid cross-talk among vacuum tubes biased by the same circuit, any residual AC signal traversing the grid leak resistors has to be shorted to ground. This is accomplished by the decoupling capacitors C_d connected between the grid leaks and the ground.

Negative voltage is obtained by connecting the positive end of the rectifier to ground. In this way, the other end of the rectifier has a voltage negative with respect to ground (and to the cathode) and can be used to provide the negative bias voltage. As usual, a reservoir capacitor and a smoothing filter follow the rectifier to produce a stable DC voltage. The voltage divider, composed of resistors R_1 and R_2, brings the voltage, exiting from the smoothing filter, to the needed grid bias voltage $-V_g$. The bias voltage goes to the grid, of the two vacuum tubes of the push-pull stage, through the two grid leak resistors R_l.

Note however that, the same negative grid bias circuit is generally used to provide bias voltage to all the vacuum tubes of all the push-pull stages. Using this schema in a stereo amplifier, the residual input signal seen at the grid of a channel can go to the grid of the other channel, through the grid leak resistors, and produce some unwanted cross-talk among the different channels. In order to eliminate this problem, the residual signal seen at the end of the grid leak resistors, should be shorted to ground. This task, as discussed in Section 3.6.1, is accomplished by the decoupling capacitors C_d.

Since, in normal operation, no current goes through the grid and the grid leak resistors, current mainly flows through the voltage divider, used to set the correct grid bias voltage. The voltage divider can be designed using appropriate large values of the resistors, to minimize the current and simplifying the job of the step-down transformer, the rectifier, the reservoir capacitor, and the smoothing filter. However, remember that, as we discussed in Section 4.1.1, vacuum tube datasheets specify a maximum value for the resistance between the grid and the cathode, to avoid the thermal runaway problem. With fixed grid bias, given that the cathode is at ground level, the resistance between grid and cathode is the sum of the grid stopper, grid leak, and resistor R_2 of the grid bias voltage divider. Therefore, there is a limitation on the value of the resistors that can be used in this voltage divider.

Example 25: Power supply design for fixed grid bias

Suppose we use a step-down transformer that provides 100V output, has a primary resistance of 4 Ohms, and has a secondary resistance of 20 Ohms. Suppose also we use a filter resistor R_{flt}=32K Ohms and the total resistance wanted in the voltage divider is 35K Ohms, so that the total load seen by the rectifier is 67K Ohms. If we chose a reservoir capacitor C_R=1 μF, and a capacitor filter C_{flt}=1μF, according to what we discussed in Sections 5.1.5, 5.1.6, and 5.1.8, we can conclude that the output of the rectifier is -135.7V and that the output of the smoothing filter is -70V. The ripple voltage at the rectifier output is around 5V and the ripple voltage at the smoothing filter output is around 0.25V.

Suppose now we want to set the grid bias voltage to -40V. Using the voltage divider equation, we obtain that we can set the resistors of the voltage divider to R_1=15K Ohms and R_2=20K Ohms. The value of R_2 has to be added to the grid leak and to the grid stopper to check that the maximum grid to cathode resistance is not exceeded. Elsewhere smaller values of R_1 and R_2 have to be chosen.

Note also that, we take the grid bias at the voltage divider. Therefore, the ripple voltage arriving at the grid will be reduced as well by the voltage divider, and will be around 0.14V. Suppose we are using EL34 vacuum tubes, since the grid voltage peak of an EL34 vacuum tube is around 35V, which is much larger than 0.14V, we can safely accept this ripple voltage at the grids, in a push-pull configuration.

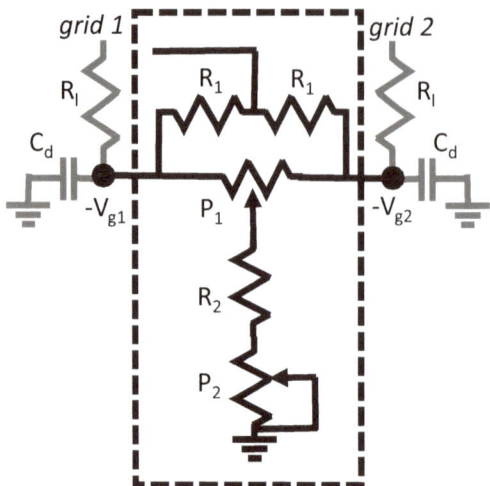

Figure 46: Improved schema for supplying the grid bias voltage
The basic voltage divider used in Figure 45, to obtain $-V_g$, is refined here to allow fine-tuning of the grid bias voltage for each vacuum tube. Potentiometer P_2 allows increasing or reducing the grid bias voltage of both vacuum tubes, to choose the wanted operating point. Potentiometer P_1, with the help of the two resistors R_1, allows fine balancing the grid bias in the two vacuum tubes of a push-pull amplifier, to have exactly the same operating point in both tubes.

5.3.1 Fine tuning the grid bias

The voltage divider, in the dashed box in Figure 45, provides the same grid bias voltage $-V_g$ to all vacuum tubes connected to it. However, pairs of vacuum tubes, even matched pairs, have some small differences and react differently under the same conditions. For instance, even if two vacuum tubes have the same anode voltage, and the same grid bias, they can conduct slightly differently and can stay on different operating points. Two not perfectly paired vacuum tubes can compromise the benefits of the push-pull configuration. In addition, suppose we also want to be able to vary the grid bias voltage in order to choose different operating points, to obtain the best sonic performance, or even to choose our preferred amplifier class (A, AB, or B). In all these cases, we require the capability to fine-tune the bias voltage, to provide each different vacuum tube with its needed bias voltage.

Figure 46 shows a modified schema of the basic voltage divider. At the bottom of the figure we can see that there is a potentiometer P_2 between R_2 and ground. The wiper terminal of P_2 is connected directly to one of the other two terminals. The position of the potentiometer modifies the resistance between its two ends. Higher resistance produces a more negative voltage, and vice versa. In this way, it is possible to increase or reduce the bias voltage of both vacuum tubes, to choose the wanted operating point. Note that if, for some reasons, the wiper fails to be in contact with the carbon track of the potentiometer, the potentiometer gives its full resistance, pushing the bias at the more negative possible voltage. In this way, in case of failure of the potentiometer, the vacuum tubes are simply cut-off, without damaging them. Other designs, in case of failure of the potentiometer, can leave the grids floating or, even worse, to ground level, by damaging the vacuum tubes.

Example 26: Setting the grid bias voltage

In Example 25 we chose a reference grid bias voltage of -40V. Suppose we want to be able to fine-tune the grid bias in a range of approximately +/- 50% of the reference. This can be obtained by choosing R_2=6K Ohms and P_2=100K. In this way, we can vary the grid bias roughly in a range from -20V to -60V.

The new schema in Figure 46 also uses the potentiometer P_1, along with the two resistors R_1, connected at the two terminal ends of the potentiometer P_1 itself. This schema allows balancing the bias between the two vacuum tubes of a push-pull amplifier, to set the same operating point, in case they react slightly differently at the same grid bias voltage. The two resistors, along with the potentiometer, act as two parallel variable voltage dividers. For instance, when the potentiometer shaft is positioned more on the left, the voltage divider on the left has the resistance between the terminal and the wiper reduced. Therefore, the grid bias voltage on the left will be less negative (closer to zero), while the one on the right will be more negative. Also in this case, if the potentiometer fails, for instance if the wiper fails to be in contact with the carbon track, the two grids are put at a more negative voltage, so cutting the vacuum tubes off and avoiding to damage them.

Example 27: Balancing the grid bias voltage of two vacuum tubes in push-pull.

Suppose we want to be able to set exactly the same operating point, for two vacuum tubes in a push-pull amplifier. Suppose that, in order to do that, we want to fine-tune their grid bias voltage in a range of approximately +/- 10% of the reference grid bias chosen in Example 25. This can be obtained by setting the two resistors R_1=33K Ohms and the potentiometer P_1=10k. In this way, for instance, if P_2 is set to provide us with a grid bias voltage of -40V, we can use P_1 to balance the grid bias of the two vacuum tubes, approximatively, in a range from -36V to -44V.

Note that if the circuit for grid bias fine tuning is used to in a stereo amplifier, the voltage divider, composed of the two parallel resistors R_1 and the potentiometer P_1, should be duplicated. Both voltage dividers should be connected to R_2 through their potentiometer P_1. In addition, the values of R_1 and P_1 should be doubled as well, since they work in parallel.

5.3.2 Probing the grid bias current

It is easy to tune the bias voltage for each vacuum tube of the push-pull stage, using the circuit shown in Figure 46. However, in order to set the correct operating point, we need to know the bias current flowing through the vacuum tubes. The bias current can be easily measured by connecting the cathodes of the vacuum tubes to ground through two resistors, indicated as R_{p1} and R_{p2} in Figure 47. The resistance of the two resistors should be very small, in order not to introduce practically any cathode voltage elevation, and local negative feedback. For instance, values of 1 Ohm or at most 10 Ohms are generally used. Some probe pins, possibly accessible without opening the amplifier chassis, are connected to the terminals of the resistors, as depicted by the pins A, B, and C, in the figure.

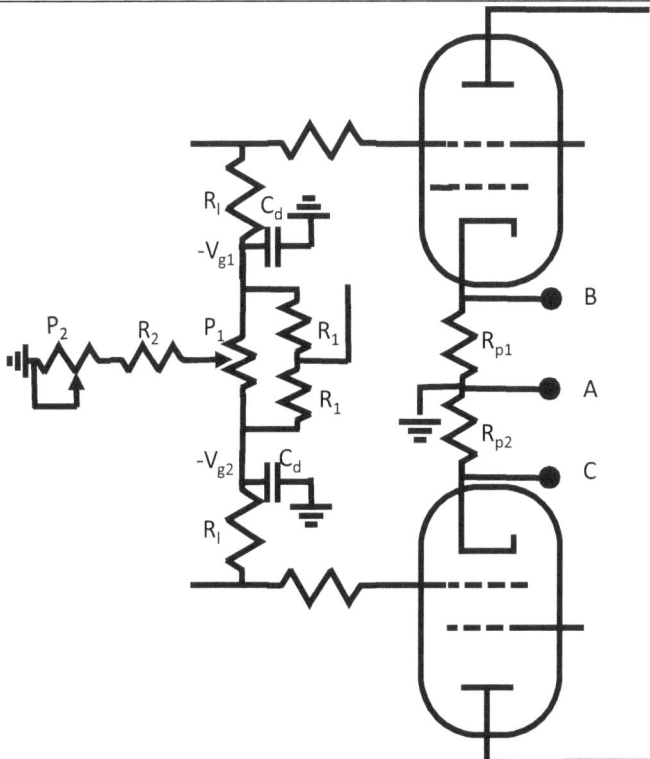

Figure 47: Probing the grid bias current
In order to be able to measure the bias current, when operating on the potentiometers of the improved grid bias circuit, the cathodes of the vacuum tubes are connected to ground through two very small resistors R_{p1} and R_{p2}. The bias current can be computed by measuring the voltage between pin A, connected to ground, and the other two pins A or B, and then using the Ohm's law.

The bias current can be obtained by measuring the voltage between pin A, connected to ground, and pin B or pin C, and by using the Ohm's law to calculate the current.

Example 28: Measuring the bias current

Suppose $R_{p1} = R_{p2} = 10$ Ohms. If we measure 0.4V between A and B, and 0,35V between A and C, then the bias current of one vacuum tube is 0.4/10=40 mA, and the bias current of the other is 0.35/10=35 mA.

Fine-tuning starts by connecting a voltmeter to either A and B, or A and C and then turning the potentiometer P_2 until the wanted bias current is measured. After this, the voltmeter must be connected to B and C, and the potentiometer P_1 turned until the measured voltage is zero, that is B and C are at the same potential. Then, the voltmeter is connected again to either A and B, or A and C and the two steps are repeated until both tubes have the wanted identical bias current.

Chapter 6:
Step by step design of a push-pull tube amplifier

In this chapter, we use all notions introduced previously to design, from scratch, a stereo push-pull tube amplifier. As before, we will proceed backwards. We start from the output stage, then the splitter and input stages. Finally, we deign the power supply.

We will design a hi-fi amplifier that provides roughly 12 Watts RMS, using EL84 power vacuum tubes in the output stage. The same calculations and schemas can be used, in a similar way, to design an amplifier with different vacuum tubes and different output power.

6.1 Design of the power stage

We adopt a push-pull configuration in ultra-linear mode. Ultra-linear mode and push-pull configurations were introduced respectively in Section 2.2.4 and Section 4.1.4. The overall schema of the power stage, including values of all components, is given in Figure 48. Let us see how these values were chosen.

EL84 datasheet specifies that anode and screen voltage can be set to 300V and max anode power dissipation is 12 Watts.

Figure 49 shows the **anode characteristic graph** of an EL84 vacuum tube, when anode voltage is 300V and ultra-linear configuration is obtained feeding the screen with 40% of the primary turns.

An **output transformer** for push-pull operation, with reactive load of 8K Ohms, provides each vacuum tube with a load of 4K Ohms, when operating in Class A. By setting the anode bias current to 28mA, we obtain the **loadline** depicted by the red line, which stays below the 12 Watts max allowed anode power dissipation.

From Figure 49 we can see that the **grid voltage bias**, needed to obtain an anode bias current of 28mA, is around -12V.

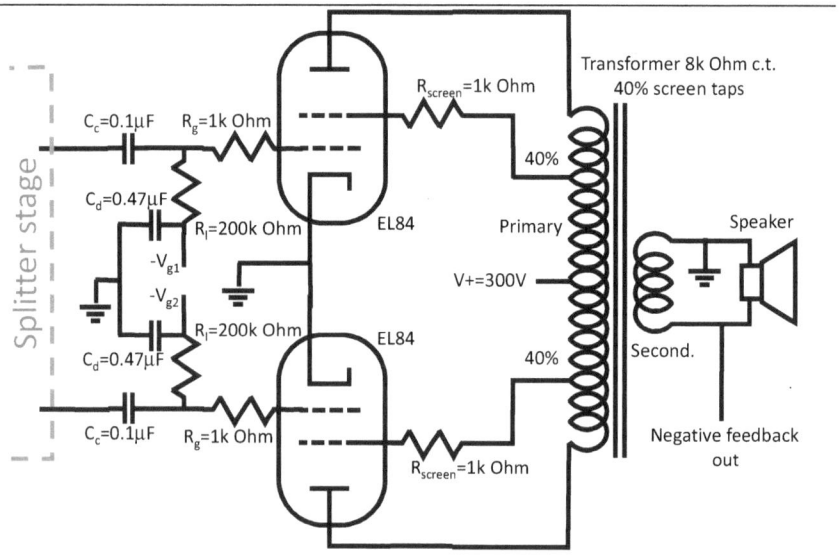

Figure 48: Schema of the amplifier output stage designed around EL84 tubes.

The amplifier operates in Class AB. More specifically, it operates in Class A when the anode voltage is in the range from 190V to 410V. When out of this range, as explained in Section 4.1.6, one of the two vacuum tubes quits conducting and the other one sees just one fourth of the whole transformer impedance. In fact, below 190V the reactive load seen by the conducting vacuum tube is 2K Ohms and the slope of the load line increases accordingly.

With this configuration, the **maximum allowed peak amplitude of the input signal** is 12V. In fact, when the input voltage is 12V, since the input signal is added to the grid voltage bias, the grid voltage becomes zero. Under these conditions, the vacuum tube saturates, since it has maximum conduction, and anode voltage is minimum.

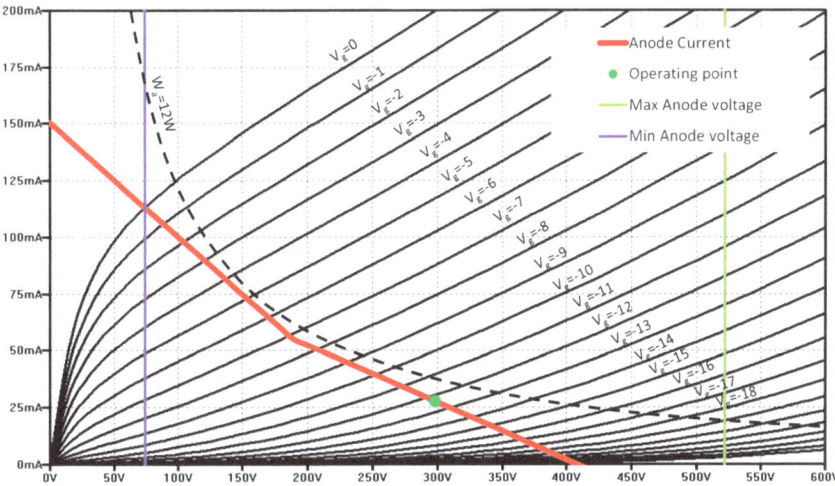

Figure 49: Average anode characteristic graph and loadline of EL84 in ultra-linear configuration.
The figure shows the anode characteristic graph of an EL84 vacuum tube operating in ultra-linear configuration. Anode voltage is set to 300V and screen is fed with 40% of primary turns. Red line represents the push-pull loadline with an 8K Ohms push-pull transformer and bias current set to 28mA.

The **grid stopper resistor** R_g, generally used with an EL84 vacuum tube, is 1K Ohms. We also use a 1K Ohms **screen stopper resistor** R_{screen} to connect the EL84 screens to the ultra-linear taps of the transformer. EL84 datasheet specifies that maximum grid to cathode resistance, with fixed bias, is 300K Ohms. Accordingly, we can use a **grid leak resistor** R_l of 200K Ohms to give some margin to the resistance of the bias voltage circuit.

Using the calculation given in Example 6 we can determine the value of the **inter-stage coupling capacitor** C_s, between the phase splitter and the output stage. To obtain a high pass filter at 7Hz, with a grid leak resistor of 200K Ohms, we set C_c to 0.1µF.

The **decoupling capacitor** C_d, to prevent residual signal at the grid leak resistor to go to the other vacuum tubes when fixed bias is used, can be computed according to Example 3. Suppose that we want to short to ground all residual signals with a frequency above 2 Hz, and that the only resistance seen by the capacitor is the 200K Ohms grid leak itself. In this case, a C_d=0.47µF accomplishes to this task. Note also that, since the grid bias voltage is provided using a voltage divider, as discussed in Section 5.3, part of the residual signal at the grid leak resistor is also shorted to ground by this voltage divider further reducing the cut-off frequency of the low pass filter.

6.1.1 Maximum output power estimation and voltage gain of the power stage

A precise estimation of the maximum output power of the power stage is difficult. It is affected by the non-linearity of the vacuum tubes and the impedance variation of the output transformer in class AB operation. However, a quite accurate estimation can be obtained as follows. From the loadline drawn on the average anode characteristic graph in Figure 49 we can see that when the inputs signal goes from 0 to 12V, the anode voltage of one vacuum tube goes from the 300V quiescent condition to 75V. The peak of the voltage variation at the anode of the vacuum tube is 300V-75V=225V. In a push-pull amplifier, when one tube is at the highest peak, the other is at the lowest peak. When the anode voltages of the two vacuum tubes are given to the two ends of the output transformer, the peak of the voltage variation is therefore doubled, and the peak voltage seen by the transformer is 2·225V=450V. This expressed in RMS gives 450V*0.707=318.15V. At this point, knowing the reactive impedance of the transformer, we are able to estimate the expected maximum RMS power seen at the primary of the transformer as $P=V^2/R=318.15^2/8K=12.7$ Watts RMS. In an ideal output transformer, this power is entirely transferred to the output and represents the maximum **RMS output power**.

With this, we are also able to estimate the **voltage gain** after the output transformer. Suppose that we use 8 Ohms speakers. Given that the maximum RMS output power is 12.7 Watts, the maximum output voltage with an 8 Ohms load is

$$V_{out} = \sqrt{P \cdot R} = \sqrt{12.7 \cdot 8} = 10.1 \text{ Volts RMS}.$$

The maximum RMS input voltage that can be applied to the power stage is V_{in}=12V·0.707=8.48V. Therefore, the voltage gain of the power stage, after the output transformer, is A=V_{out}/V_{in}= 10.1/8.48=1.19. Expressed in dB, we have A_{db}=20·log(1.19)=1,51 dB. We will see that the value of the voltage gain will be useful when designing the global feedback circuit.

6.2 Design of the input and phase splitter stage

In previous section, we noted that the maximum possible peak amplitude of the input signal to the power stage is 12V. With such a signal, the amplifier delivers its maximum power. Accordingly, when designing the input and phase splitter stages we should guarantee that they are actually able to provide the power stage with this voltage, in correspondence of the input signal received by the input stage. Many audio sources provide a signal with a maximum amplitude peak around 1V, and some CD players and DAC can even reach 2.5V or 3V.

To have maximum output power with a 1V peak amplitude input signal, we need a combination of input and splitter stage with a gain at least of 12 (or 21.58 dB).

12AX7 vacuum tubes are good candidates for this job. In fact, in Section 3.4, we said that 12AX7 vacuum tubes can provide a gain greater than 60 (around 35.56 dB). This is much more than what we need. However, this also gives a lot of space to design the negative feedback circuit of the amplifier, which, on one hand reduces harmonic distortion, on the other hand reduces the gain of the amplifier.

We will use a directly coupled concertina configuration, as discussed in Section 4.3.1. In this configuration there is no coupling capacitor between input and splitter stage, and the anode of the input stage also provides the grid bias voltage to the concertina vacuum tube.

The complete schema and the values of the various components is given in Figure 50. Let us see how these values were chosen.

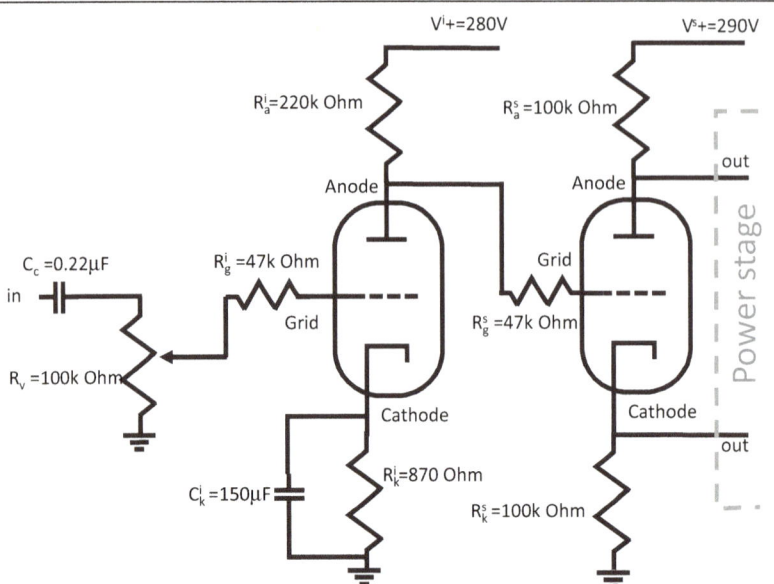

Figure 50: Schema and component values of the directly coupled input and concertina stages

6.2.1 Concertina phase splitter

The **high-tension voltage of the concertina** is set to 290V, **the cathode resistor** R_k^s and **anode resistor** R_a^s have both a resistance of 100K Ohms. This gives a total load of 200K Ohms and produces the **DC loadline** depicted by the red line in Figure 51. Under AC operation, the two grid leak resistors of the power stage are in parallel with the anode and cathode resistors of the concertina vacuum tube. The AC loads seen respectively at the anode and the cathode becomes 100K·200K/(100K+200K)=67K Ohms, for a total of 134K Ohms. Suppose we set the **operating point of the concertina** at 94.5V and 0.98 mA, as depicted by the red spot. This, along with the AC load, gives the **AC loadline** depicted by the green line in the figure. The concertina will operate along this loadline.

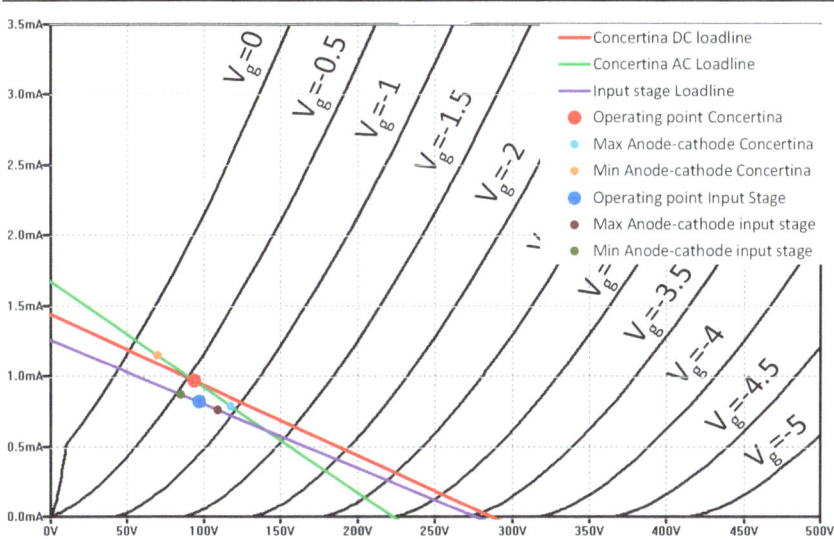

Figure 51: Loadlines, operating points, and operating ranges for directly coupled concertina with 12AX7.
The figure shows the anode characteristic graph of a 12Ax7 vacuum tube. Red line is the DC loadline of the concertina phase splitter. Green line is its AC loadline. Pink line is the loadline of the input stage. Operating point, as well as operating intervals are reported.

We already mentioned that the maximum possible peak amplitude of the input signal to the power stage is 12V. Anode and cathode of the concertina phase splitter should be able to move from their quiescent voltage of such quantity. The voltage indicated in the anode characteristic graph, in Figure 51, represents the voltage measured between the anode and the cathode of the vacuum tube. In a concertina, when anode voltage increases, cathode voltage decreases of the same quantity. This means that when the anode voltage increases of 12V, the cathode decreases of exactly the same amount. Therefore, the maximum voltage increase measured between anode and cathode, with respect to the quiescent voltage, is 24V. The orange and cyan spots, in Figure 51, highlights the maximum and minimum anode to cathode voltage (+24V/-24V with respect to the quiescent voltage), which correspond to the 12V output signals seen at the anode and the cathode. This interval represents the operating range of the concertina phase splitter, when providing the power stage with the needed signal for maximum output power. We can see that the operating range is in a rather linear area and has a reasonable distance from the vacuum tube saturation, reached when V_g=0.

In Section 4.2.2, we already mentioned that the **voltage gain** of the concertina phase splitter is A=0.98. Expressed in dB we have A_{db}=20·log(0.98)=-0.18 dB.

6.2.2 Input stage for directly coupled concertina

The **high-tension voltage of the input stage** is set to 280V. With **an anode resistor** R_a^i of 220K Ohms we obtain the **loadline** depicted by the violet line in Figure 51.

In a directly coupled concertina, the anode of the input stage vacuum tube provides the grid voltage bias to the concertina. The concertina operating point depicted by the red spot in Figure 51, corresponds to a grid to cathode voltage of approximatively V_g =-0.6V. The concertina quiescent current is 0.98 mA, therefore, the cathode voltage, elevated by the cathode resistor, is 100K Ohms·0.98 mA= 98V. To provide the concertina vacuum tube with the correct grid to cathode bias voltage of -0.6V, the input stage should be configured so that the anode quiescent voltage is very close to 98V-0.6V=97.4V.

According to the input stage loadline, with a grid bias voltage V_g=-0.7V we obtain an anode to cathode voltage of 96.7V and a current of approximately 0.8 mA. This can be obtained using the self-bias technique, described in Section 3.6.2, with a **cathode resistor** R_k^i of 0.7V/0.8mA= 870 Ohms.

With this configuration, the anode to ground quiescent voltage of the input stage is 96.7V+0.7V=97.4V, which is what needed to provide the bias voltage to the grid of the concertina phase splitter and to use directly coupling between the two stages.

In order to eliminate the local feedback effect and the gain loss, introduced by the cathode resistor, we bypass it with a **bypass capacitor** C_k^i, of 150μF, as discussed in Section 3.6.3 and Example 5.

The grid of the 12AX7 vacuum tubes should be connected to the **grid stopper resistor** R_g^i, to produce low pass filters for very high frequencies, exploiting the Miller effect, as discussed in section 4.1.1 and other places in this book. Values around R_g^i=47K Ohms are generally used for 12AX7 vacuum tubes.

Input signal arrives to the input stage grid through the **volume potentiometer** R_v. This potentiometer also acts as a grid leak for the input stage and determines the input impedance of the amplifier. We chose a logarithmic potentiometer of 100K Ohms for this purpose. The **input coupling capacitor** C_c isolates the amplifier from possible DC voltage arriving with the input signal. It basically forms a high-pass filter with the potentiometer. A capacitance of 0.22µF, with a 100K Ohms resistance provides a 7Hz cut-off frequency.

The concertina phase splitter has a gain near to unity, as discussed in Section 4.2.2. Therefore, also the peak amplitude of the output signal of the input stage should be around 12V, as required by the power stage. The green and brown spots, in Figure 51, show the operating range of the input stage, when the anode voltage signal (the output signal of this stage) swings from -12 to +12V, with respect to the quiescent condition. We can see that the operating range is again in a linear area and far from the vacuum tube saturation (V_g=0).

The **voltage gain** of the input stage can be computed as discussed in Section 3.4 as

$$A = \mu \frac{R_L}{R_L + r_a} = 100 \frac{220k}{220k + 75k} = 74.6 .$$

Expressed in dB, we have A_{db}=20·log(74.6)=37.45 dB.

As we said before, this is much higher than what required and gives us space for conveniently setting the global negative feedback circuit, as discussed in next section.

6.3 Global Negative Feedback

Previously, we determined that the gain of the input stage is 74.6, or equivalently 37.45 dB. We also discussed that the peak amplitude of the input signal that can be given to the power stage is 12V, and that the concertina phase splitter practically forwards, to the power stage, almost the entire output signal produced by the input stage.

Under these conditions, the input stage produces a 12V peak amplitude output signal when the peak amplitude of the input signal is just

V_{in}=12V/74.6=0.16V. Generally, input sources produce a much larger signal. As we already said, some input sources, as for instance CD players or DACs, can even provide a 2.5V or 3V peak signal.

For instance, if we want to produce the maximum output signal, with a 2.5V peak signal, the gain needed at the input stage is just A=12V/2.5V=4.8, or equivalently 13.62 dB. Accordingly, we can apply a global negative feedback up to fb_{db}=37.45-13.62=23.83 dB and we are still able to have the maximum output power.

If we want to obtain maximum output power also with sources providing less input voltage, as for instance just 1V peak, the needed gain of the input stage is A=12V/1V=4.8 or equivalently 21.58 dB. This gives us space for a **negative feedback** fb_{db}=37.45-21.58 =15.87 dB.

The amount of negative feedback is determined by the feedback factor β, as discussed in Section 4.4. The value of β can be computed knowing the gain of the amplifier, when no negative feedback is used. This, in turn, can be determined using the gain of the individual stages, which we have computed in previous sections. The voltage gain of the amplifier with no negative feedback is A=74.6·0.98·1.19=87, or alternatively A_{db}=37.45-0.18+1.51=38.78 dB.

As discussed in Section 4.4.1, the amount of feedback is $fb_{dB} = 20 \cdot \log(1 + A\beta)$. When we apply a feedback fb_{db}=15.87 dB, to provide maximum power with 1V peak input signal with an amplifier gain A=87, we have that the **feedback factor** β is

$$\beta = \frac{10^{fb_{db}/20} - 1}{A} = \frac{10^{15.87/20} - 1}{87} = \frac{5.22}{87} = 0.06.$$

To inject the negative feedback, at the cathode of the input stage vacuum tube, we split the cathode resistor, used for cathode biasing, into two resistors R_k and R_2. The negative feedback is injected between the two resistors through the resistor R_1, as depicted in Figure 52. Remember that $\beta = R_2/(R_1 + R_2)$. Therefore, if **resistor** R_2 is set to 150 Ohms, β=0.06 can be approximatively obtained by choosing the **feedback resistor** R_1=2.2K Ohms. The **cathode resistor** R_k, which was previously set to 870 Ohms, should now be reduced, to take into account the 150 Ohms used for R_2. The cathode resistor should be 870 Ohms–150 Ohms=720 Ohms. The closest standard is 680 Ohms, which is near to the computed value.

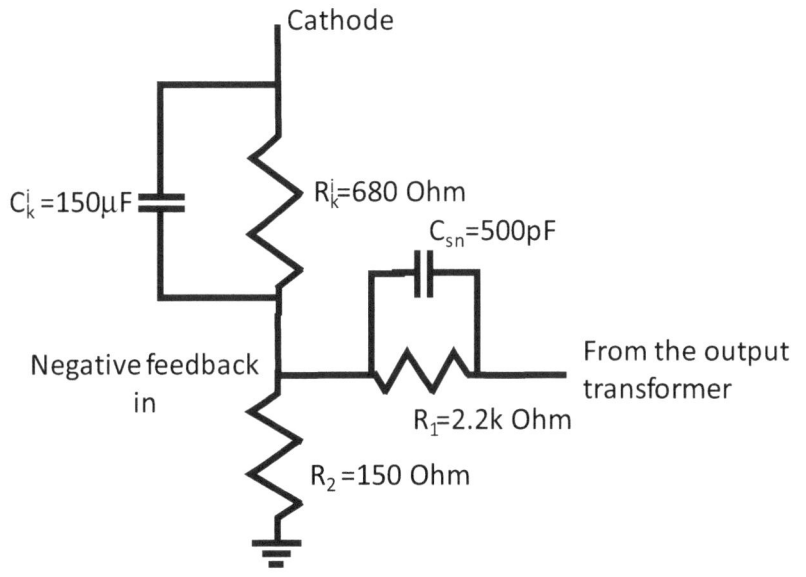

Figure 52: Schema and component values for global negative feedback

The attenuation to high frequencies, due to the low-pass filter composed of the grid stopper resistors and the Miller capacitance, brings the feedback fb_{db} below 6dB, at frequencies higher than 50KHz. With this negative feedback, the possible occurrence of a 180° degrees phase shift (which generally does not occur at lower frequencies) is not harmful and the amplifier should be already stable, as discussed in Section 4.4.3. However, we can reduce radio frequency interference and further increase stability to the circuit by using a 500pF **capacitor** in parallel to R_2. This capacitor has the role of reducing the phase shift of the negative feedback signal, in the range from 20KHz to 10MHz. It also increases the negative feedback effect in that range, by attenuating any high frequency intercepted by the input of the amplifier.

Figure 52 shows the schematics and the value of the various components needed for global negative feedback circuit.

6.4 Design of the amplifier power supply unit for the amplifier stages

The schema of the power supply unit for the stages of the amplifier is given in Figure 53. As discussed in Section 5.1, it is composed of a power supply transformer, a rectifier with its reservoir capacitor, and two sequences of RC smoothing filters, one per each channel of the stereo amplifier.

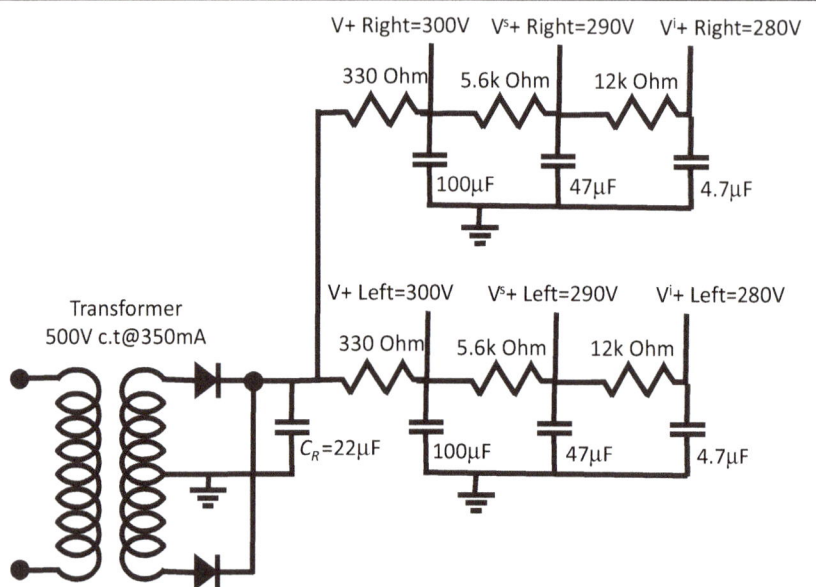

Figure 53: Power supply unit for the stages of the amplifier

We start computing the voltage delivered by the rectifier via the reservoir capacitor, and its ripple voltage. We estimate the rating of the power transformer. Then, we compute the resistance of the smoothing filters. Finally, we compute the capacitances of the smoothing filters.

6.4.1 Voltage delivered by the rectifier and ripple voltage

The voltage produced by the rectifier via the reservoir capacitor, as discussed in Sections 5.1.5, depends on

- the type of rectifier,
- the mains frequency,
- the output voltage of the power transformer,

- the output impedance of the power transformer,
- the load seen by the reservoir capacitor,
- the capacitance of the reservoir capacitor

We will use **a full wave rectifier** and consider a **mains frequency** f_M=50Hz. **A 500V centre tapped power transformer** has a peak voltage V_s^p =250V·1.414=353.5V. Suppose the **output impedance of the power transformer** is around R_s=27 Ohms, which is a realistic value.

The load seen by the reservoir capacitor is the impedance of the power supply filter plus the impedance of the entire amplifier. The impedance of the amplifier can be estimated by using the Ohm's law, by dividing the voltage given to the first powered stage, which is the power stage, by the current absorbed by all stages. The quiescent current absorbed by the concertina and by the input stage was already discussed before and is respectively 0.98mA and 0.8mA. The quiescent current of the power stage is 28mA in each tube, which makes 56mA. When working in class A, the current absorbed by the power stage is always 56mA, since when one tubes conducts less the other conducts more. When delivering maximum output power, working in Class AB, the maximum absorbed current arrives up to 115mA. However, we will use Class A conditions to compute the expected DC voltage produced by the reservoir capacitor. So, we use 56mA as a reference current for the power stage. This implies that when operating at maximum output power, the voltage delivered by the reservoir capacitor will be slightly lower. The total current absorbed by the amplifier is therefore 56mA+0.8mA+0.98mA=57.78mA. We have two channels so the total current should be doubled, obtaining I_L=115,56mA. The voltage needed at the first powered stage, which is the power stage, is 300V. Therefore, the impedance of the amplifier is R_L=300V/115.56mA=2.6k Ohms. We said that the total load seen by the reservoir capacitor is the impedance of the amplifier plus the impedance of the smoothing filter of the power stage. We will see that, given that the voltage delivered by the reservoir capacitor is close to what needed by the power stage, the impedance of the power stage smoothing filter is much smaller than the impedance of the amplifier, so we can ignore it and use only the impedance of the amplifier.

We now have all ingredients to estimate the voltage delivered by the reservoir capacitor. Suppose we use **reservoir capacitor** with a $C_R=22\mu F$. The ratio R_S/R_L is 1.01% and $2\pi f_M C_R R_L=18.4$. According to the graph in Figure 39 we have a percentage V_{dc}/V_S^p around 91%. **The estimated DC output voltage at the reservoir capacitor** is $V_{dc}=352.5V\cdot 91\%\approx 321V$.

Similarly, using the graph in Figure 40 we can determine the expected **ripple voltage**, ad discussed in Section 5.1.6. The green line, corresponding to $R_S/R_L=1\%$, in correspondence of $2\pi f_M C_R R_L=18.4$, gives a ratio V_{ripple}/V_{dc} around 4%. Therefore, the expected voltage ripple is $V_{ripple}=321\cdot 4\%=12.84V$.

6.4.2 Rating of the power transformer

Using the parameters computed in previous section we can also compute the **RMS current of the power transformer secondary**, with the help of Figure 41. The red plot corresponding to $R_S/R_L=1\%$, in correspondence of $2\pi f_M C_R R_L=18.4$, gives a ratio I_S/I_L around 1.6. Therefore, the expected RMS current of the power transformer secondary is $I_S=115.56mA\cdot 1.6=184.8mA$. To guarantee safe operation we can chose a centre tapped 500V power transformer rated for a 350mA RMS current, or simply 500V C.T. @350mA.

6.4.3 Resistors of the smoothing filters

The two chain of filters, for the left and right channels of the amplifier, are identical, so we just need to compute values for one chain. Computation of the components of a smoothing filter was discussed in Section 5.1.8. We first compute the value of the resistors of the smoothing filters, going backward, from the last filter of the chain, which is the filter for the input stage.

In Section 6.2.2, we determined that the anode voltage of the 12AX7 vacuum tube of the input stage is 280V. Its quiescent current is 0.8mA. When delivering maximum power, the current varies +/- 0.05mA from the quiescent current and has little impact, so we can safely use the quiescent current in our computation. We also determined that the voltage required by the anode of the 12AX7 vacuum tube of the next stage, the concertina phase splitter, is 290V. As explained in Example 21, the **resistor of the smoothing filter between the concertina and the**

input stage has to produce a voltage drop of 10V. Given that the current traversing the resistor is 0.8mA, we have that its resistance must be 10V/0.8mA=12.5K Ohms. The closest standard is 12K Ohms, which is close to the needed value. The power dissipated by the resistor is 10V·0.8mA=0.008W. Therefore, we can use a resistor rated for 0.5W.

The voltage of the anodes of the EL84 vacuum tubes is 300V. The current absorbed by the concertina at its quiescent state is 0.98mA. At maximum power, the current varies +/- 0.18mA, with respect to the quiescent state. Also in this case, the variation does not affect significantly our computation, and we use the quiescent current as reference. To obtain the needed 290V at the concertina, the **resistor of the smoothing filter between the power stage and the concertina** has to produce a voltage drop of 10V. The current traversing this resistor is the current absorbed by the input stage and that absorbed by the concertina, that is 0.8mA+0.98mA=1.78mA. The value of the resistor is 10V/1.78mA=5.6K. The power dissipated by the resistor is 0.018W. Also in this case we case we can use one rated for 0.5W.

Finally, we compute the **resistor of the smoothing filter of the power stage**. It should drop 21V, from the 321V delivered by the reservoir capacitor, to obtain the wanted 300V. The current traversing the resistor is the sum of the current absorbed by the power stage, the concertina, and the input stage, that is 56mA+0.8mA+0.98mA=58mA. The value of the needed resistance is 21V/58mA=362 Ohms. The closest standard is 330 Ohms. The power dissipated by the resistor is 1.34W, so it is convenient to choose one rated for 5W. The computed resistance is much smaller than the impedance of the amplifier. Therefore, as we said before, adding it, to exactly compute the total load seen by the reservoir capacitor, does not affect significantly our estimation.

6.4.4 Capacitors of the smoothing filters

We can now choose the values of the capacitors of the smoothing filters. In section 6.4.1 we computed that the ripple voltage after the reservoir capacitor is 12.84V. The DC voltage of the power stage is 300V, so this ripple voltage is around 4% of DC voltage. Recommended ripple voltage in a push-pull stage is around 0.5% to 2% (see Section 5.1.8) so we need to significantly attenuate it. If the **capacitor of the smoothing filter of** the power stage is 100µF, according to Section 5.1.8, we have that the voltage ripple is

$$V^o_{ripple} = V^i_{ripple} \frac{1}{\sqrt{1+(2\pi \cdot R_{flt} \cdot C_{flt} \cdot 2f_M)^2}} =$$

$$= 12.84V \frac{1}{\sqrt{1+(2\pi \cdot 330 Ohm \cdot 100\mu F \cdot 100 Hz)^2}} = 0.62V.$$

This corresponds to 0.2% of the power stage DC voltage, so a very good value.

Following the same procedure, we chose the value of the **capacitor of the smoothing filter of the phase splitter stage.** With a capacitor of 47µF, we have that

$$V^o_{ripple} = 0.62V \frac{1}{\sqrt{1+(2\pi \cdot 5.6 KOhm \cdot 47\mu F \cdot 100 Hz)^2}} = 0.0037V$$

This corresponds to 0.001% of the phase splitter DC voltage, which is 290V, and lower than acceptable values, which are around 0.01% to 0.05%.

Finally, we set the value of the **capacitor of the smoothing filter of the input stage.** With a capacitor of 4.7µF, we have that

$$V^o_{ripple} = 0.003V \frac{1}{\sqrt{1+(2\pi \cdot 12 KOhm \cdot 4.7\mu F \cdot 100 Hz)^2}} = 0.000085V$$

The DC voltage of the input stage is 280V. The ripple voltage corresponds to 0.00003% of the DC voltage. Ripple voltage percentage is, also in this case, smaller than safe values, which are around 0.001% to 0.002%.

6.5 Design of the vacuum tube filament power supply

We need to provide power to the filaments of the four EL84 and to the filaments of the two 12AX7, which in turn contain each two vacuum tubes. According to their datasheets, current absorbed by the filament of one EL84 is 0.76A. Current absorbed by the filament of one section of a 12AX7 is 0.15A. Voltage needed by all these filaments is 6.3V. Using a power transformer providing the needed 6.3V, the total current absorbed by the filaments is 4·(0.76+0.15)=3.64A. To stay on the safe side, **the transformer for the power supply of the filaments** can be 6.3V@6A.

The schema of the power supply for the filaments is given in Figure 54.

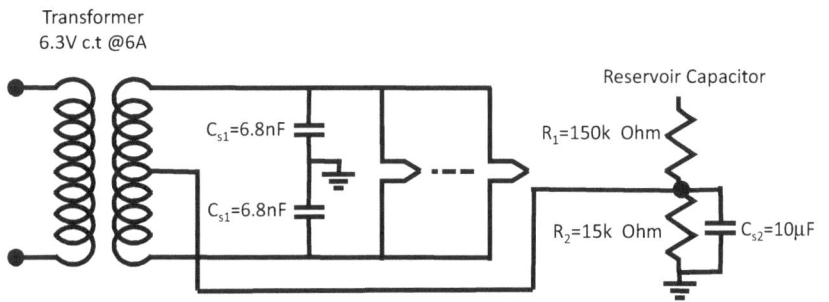

Figure 54: Power supply for the filaments

Some other components need to be added to the bare power transformer. As discussed in Section 5.2.1 the filament circuit cannot be floating with respect to the other power supplies, and a reference voltage needs to be set. EL84 datasheets specifies that voltage between filament and cathode should be lower than 100V. Similarly, 12AX7 datasheet specifies that voltage between filament and cathode should be between +/-180V. Cathode voltage of the input stage is practically at ground voltage. Cathode voltage of the power stage is also practically at ground, if fixed bias is used. However, cathode voltage of the concertina splitter is 98V, as discussed in Section 6.2.1. If we elevate the reference voltage of the filaments to 30V we are on the safe side. **Voltage elevation can be obtained with a voltage divider**, connected to the transformer centre tap, starting from the DC provided by the reservoir capacitor. In Section 6.4.1 we determined that the expected voltage at the reservoir capacitor is 321V. To obtain the needed 30V, the relationship between the resistors of the voltage divider should be

$$\frac{R_2}{R_1 + R_2} = \frac{30V}{321V} \approx 0.09.$$

When choosing these resistors, we should also take into account the maximum impedance from the filament to the cathode. EL84 datasheets specifies that impedance between filament and cathode, which corresponds to R_2, should be lower than 20K Ohms. 12AX7 tolerates much higher impedance between filament and cathode, which can reach 150K Ohms. To stay below the needed thresholds, we chose R_2=15K Ohms, and correspondingly R_1= 150K Ohms.

Resistor R_2 is bypassed by a 10µF capacitor, to attenuate ripple coming from the reservoir capacitor. We also use two small 6.8nF capacitors, from the two transformer ends to ground, to short to ground high frequency interferences intercepted by the circuit. Finally, if we do not have a centre tap transformer, we can create an artificial centre tap, as discussed in Section 5.2.2, by connecting the voltage divider to two 220 Ohms resistors, connected in turn to the two transformer ends.

6.6 Design of the fixed grid bias power supply

Power supply for the grid bias needs to produce DC voltage. Therefore, the power supply is, as usual, composed of a transformer, a rectifier, a reservoir capacitor, a smoothing filter. Finally, there is a voltage divider to obtain the needed voltage. Remember that grid voltage bias is negative with respect to the cathode, which is at ground level. In order to obtain negative voltage, the positive of this power supply is connected to ground and the grid voltage bias is taken from the negative. The complete schema of our power supply for the grid bias is given in Figure 55. Let us analyse how the values of the various components are obtained.

We first compute a simple voltage divider, as the one in Figure 45, then we refine it to obtain the improved circuit shown in Figure 55. In Section 6.1 we determined that the needed grid voltage bias is -12V. Let us assume that the DC voltage obtained after the smoothing filter is -40V. In order to obtain the wanted -12V, the relationship between the two resistors of the voltage divider is as follows:

$$\frac{R_2}{R_1 + R_2} = \frac{12V}{40V} = 0.3.$$

Remember also that datasheets specify the maximum allowed impedance between the grid and the cathode of a vacuum tube. In case of the EL84, the maximum impedance between the grid and the cathode is 300K Ohms, when used with fixed bias. In our case, this impedance is the sum of the grid stopper (1K Ohms), the grid leak (200K Ohms), and R_2. Accordingly, the value of R_2 should be smaller than 100K Ohms. If we chose R_1=35K Ohms and R_2=15K Ohms, we respect the relationship between the resistors and the limitations on the impedance between grid and cathode as well. The total resistance of the voltage divider is 50K Ohms. With a -40V voltage, the current that flow through the voltage divider is 40V/50K Ohms=0.8mA, which correspond to 32mW dissipated. These very small values allow using a small power transformer.

We can now modify the values computed above, according to the guidelines discussed in Example 26 and Example 27, in Section 5.3.1. In addition, we have to consider that we need two different voltage dividers, for the left and right channel, so the values of R_1 and P_1 should be doubled. We obtain the circuit in Figure 55. Value of R_2 is brought to 10K Ohms and a potentiometer P_2 is connected in series to allow adjusting the grid voltage bias of 25%, roughly from -9V (when P_2 resistance is minimum) to -15V (when P_2 resistance is maximum). In place of one single resistor R_1, we use two resistors of 150K Ohms, connected to the two ends of the 10K Ohms potentiometer P_1. The circuit with the two resistors R_1 and the potentiometer P_1 is duplicated twice, for the left and right channel. This allows balancing voltage of the two vacuum tubes in the push-pull stage of 10%, roughly from -11V (when P_1 fully turned on one side) to -13V (when P_1 fully turned on the other side).

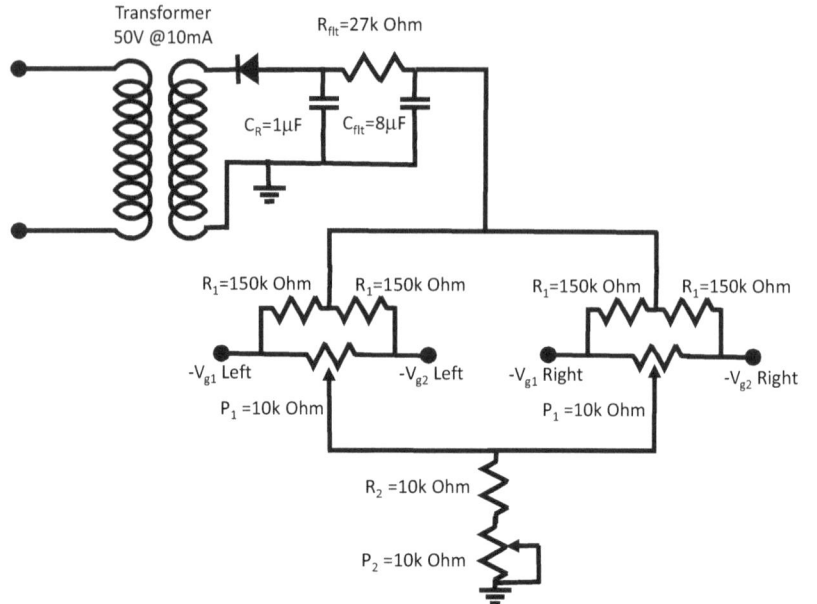

Figure 55: Power supply for the grid bias

The voltage divider absorbs very little current, and correspondingly its resistance is very high. Therefore, even with a 1µF reservoir capacitor and with a half wave rectifier, DC voltage is around 90% of RMS voltage. We can use a 50V power transformer, and we obtain 50V·1.414·90%=63,6V DC. The ripple voltage is around 6% of the DC voltage, so around 4V RMS.

We computed the voltage divider assuming to have -40V DC out of the smoothing filter. Therefore, the resistor of the smoothing filter should drop around 23V. Remember that the voltage divider absorbs 0.8mA, therefore the resistor of the smoothing filter is R_{flt}=23V/0.8ma=29 Ohms. The closest standard is 27 Ohms. Using an 8µF capacitor, we are able to reduce voltage ripple to 0.13% of the DC voltage. This, related with the -12V, corresponds to 15mV, which can be safely used in a push pull stage.

www.ingramcontent.com/pod-product-compliance
Lightning Source LLC
Chambersburg PA
CBHW040218220526
45473CB00001B/32